American Philosophical Society Library
Publication Number 12

New Perspectives on Technology and American Culture

BRUCE SINCLAIR, EDITOR

The American Philosophical Society · *Philadelphia* · *1986*

*Copyright © 1986 by American Philosophical Society
for its Library Publications Series,
Number 12*

Cover illustration: John R. Freeman's idea of what the new building at MIT should look like. Study #7, Freeman Papers, MIT Archives.

LC: 86-71782
ISBN: 0-87169-390-9

Contents

FOREWORD   vii

BRUCE SINCLAIR   1
Inventing A Genteel Tradition: MIT Crosses the River

CARROLL W. PURSELL   19
*"What the Senate is to the American Commonwealth":
A National Academy of Engineers*

JEFFREY MEIKLE   31
Materials and Metaphors: Plastics in American Culture

BRIAN HORRIGAN   49
Popular Culture and Visions of the Future in Space, 1901–2001

MICHAEL L. SMITH   69
*Back to the Future: EPCOT, Camelot, and the
History of Technology*

CONTRIBUTORS   80

Foreword

These papers were originally presented at the American Philosophical Society, in a colloquium also entitled "New Perspectives on Technology and American Culture." Two things tempted me to use such ambitious language. One was the energetic interest of the American Philosophical Society's Librarian, Edward C. Carter II, in the history of American technology, and his willingness to encourage new lines of inquiry in that field. The other was my conviction that interesting possibilities for research lay in the use of different kinds of source material and in different modes of analysis than those we usually bring to the subject.

In particular, it appeared to me that some of the brightest and most provocative work had recently come from people who wouldn't ordinarily describe themselves as historians of technology, but who nonetheless employed the insights of cultural anthropology and cultural history to reveal important, new truths about American technology. Tony Wallace's celebrated study *Rockdale* is an obvious example, but I think as well of John Kasson's *Civilizing the Machine,* or the work of Alan Trachtenberg, Warren Susman, and Jackson Lears.[1] What distinguishes the scholarship of these people is their way of seeing the problem. As a case in point, historians of technology already knew many of the details of John Kasson's story, but had never actively linked technical change, early industrialism, and republican values. So here was a new kind of thinking about familiar materials and a new application for the sources of literary and cultural history. The most telling characteristic of this literature, however, was that it pointed directly toward the relations between technology and American life; it situated technology in the largest possible context. For years, historians of technology have claimed that their subject was at the center of the American experience. Here, finally, was an approach that could test the argument.

On the face of it, one could say that technology has historically been crucial to Americans either as an engine of economic change, or as a cultural force, or for many people as both those things, closely tied together. Kurt Vonnegut remembers that in his childhood during the Great Depression of the 1930s, his family retained undiminished an almost religious faith in technology to solve America's problems.[2] I wanted the colloquium to deal with this kind of popular sentiment as a way of linking real events to general perceptions of them, and it occurred to me that might be done by joining the work of historians of American technology with the inquiry of scholars in the field of American Studies.

There is a suggestive case to be made for the idea. As an illustration, the concentration of economic power in the U.S. by the late nineteenth century ought to have rendered largely irrelevant those innocent notions of the Jacksonian era that technical advance, by providing the means for social and economic mobility, was to be the mainspring of the democratic system. Yet, like the Vonnegut family, a great many Americans continued to believe in some form of that vision. How are we to understand this apparently contradictory behavior? The best answer is surely the one that logically connects technology as material reality and technology as the agent of cultural transformations. Our contemporary dismay over consumerism, Warren Susman pointed out, does not alter the fact that for Americans of an earlier day abundance was powerfully attractive, and the inventive activity that created it also vigorously shaped their culture.[3]

I suppose I should confess, however, that the colloquium whose papers are published here did not flow quite so directly and unaided from notions of new approaches to the history of American technology. My own current research started with a similar, though more restricted idea; namely that it would be interesting to connect the existing literature on the "internal" history of the American engineering profession with an "external" study of engineering as it appeared in the general culture of the same period. But it was a particularly bright and lively session at the 1984 annual meeting of the Society for the History of Technology, entitled "Popular Culture and Technology in Twentieth-Century America," that persuaded me this combination of research styles had a wider utility.

Though none of these papers was originally designed for the colloquium, or to demonstrate the validity of a particular research approach, they ended up serving both purposes. Here is independent work in progress by historians of American technology and by scholars in American Studies, yet as Michael Smith shows in his commentary, there are a number of connections between all four papers. Those symmetries proved fruitful in the discussion that followed the oral presentation of these papers and all of us hope that in published form they may attract the consideration of other scholars interested in the subject.

It is a pleasure to acknowledge the creative participation of those who came to hear these papers last April, and the amiable assistance of the Library's staff in making that day such a pleasant occasion. Most of all, I am grateful to the Librarian for appointing me Andrew W. Mellon Senior Research Fellow in 1984–85, and to the Mellon Foundation for its support of the colloquium.

Bruce Sinclair

1. Anthony F.C. Wallace, *Rockdale* (New York: Knopf, 1978); John Kasson, *Civilizing the Machine: Technology and Republican Values in America, 1776–1900* (New York: Grossman Publishers, 1976); Alan Trachtenberg, *The Brooklyn Bridge: Fact and Symbol* (Chicago: University of Chicago Press, 1979); Warren Susman, *Culture as History: The Transformation of American Society in the Twentieth Century* (New York: Pantheon Books, 1984); T.J. Jackson Lears, *No Place of Grace: Anti-modernism and the Transformation of American Culture, 1880–1920* (New York: Pantheon Books, 1981).

2. Kurt Vonnegut, *Palm Sunday: An Autobiographical Collage* (New York: Delacorte Press, 1981).

3. Warren Susman, *Culture as History,* xxvi.

Inventing A Genteel Tradition: MIT Crosses the River

Bruce Sinclair

It seems contradictory to speak of engineering culture. For if the expression implies elevated creative accomplishment, it runs up against the antipathy of aesthetes, who argue that the finer things in life can only come from disinterested knowledge. But when the culture of engineering is defined generally to mean a certain view of the world and of the engineer's place in it, then historical opinion is against the idea, too. One recent study argues that the profession grew up in America along with corporate capitalism and, like plants that turn naturally toward the sun, engineers bent their wills to that more powerful influence. In this scenario, engineering culture is simply business culture. An alternative view is that bureaucracy gets in the way of an effective professional identity as much as business does, leaving engineers caught between the two and unable to work out for themselves a coherent form of self-realization.[1]

But maybe the truth is that an engineering ethos cannot be described simply. Besides the tension between capital's demands and the profession's ideals, there are interesting differences between engineering as business, as an art, and as science. These things have changed over time, too, and they have always been related to American culture, encompassing views about technology as diverse as those of Eugene O'Neill's play *Dynamo* and E.B. White's essay on the joys of Model T ownership.[2]

Indeed, the general culture is the place to begin the analysis. Listen, for example, to the characterization of engineers that appears in a short story published in the *Atlantic Monthly* in 1906 called "A Girl of the Engineers." Told from a feminine point of view, it deals with the special qualities needed

---

1. David F. Noble, *America by Design: Science, Technology and the Rise of Corporate Capitalism* (New York: Alfred Knopf, 1977), and Edwin T. Layton, *The Revolt of the Engineers* (Cleveland: Western Reserve University Press, 1971).

2. White's marvellous essay originally appeared in the *New Yorker* with the title "Farewell, My Lovely!" and was reprinted in *Essays of E.B. White* (New York: Harper & Row, 1977).

by women who marry engineers. Those traits were just what you would expect: the ability to put up with isolation and primitive conditions in remote engineering camps, with husbands totally absorbed by their work, and with the emotional obtuseness of these "big boy men" who, in their appealing way, required a lot of mothering.

Kate, the central character and the eldest daughter of an engineer, described those features to her brother's fiancée by way of educating her, since the brother was also an engineer:

the profession selects its own men, you know. And then out of those men some want the jobs in the cities near the crowds and the theatres and the girls; and the others—Those are the men I know; they have been trained to stand alone, to talk little, never to complain, to bear dullness and monotony, some of them are dull and monotonous themselves. But they aren't petty; and in every one of them there is a strange need that drives them out into the deserts; a craving for movement and freedom and fresh new air that nothing can kill. And oh, but I'm glad it is so. It's what keeps them young; it's what makes them strong and exciting and different; it's what makes their gentleness so wonderfully gentle; it's what makes us love them.[3]

You could imagine Gary Cooper or Jimmy Stewart playing the role, dressed in tall, laced-up field boots, riding pants, checkered shirt, and a battered felt hat. Here is a direct and straightforward man, incapable of artifice and unaccustomed to elegance—someone totally committed not only to getting the job done, but done right and on time. These are men of action who see life in simple terms; in their driving, restless spirit it is easy to recognize the mythic elements of the American male: strong silent types, builders, men who eat on the run and thrive on the challenge of their work.

But how can we account for the remarkable resonance of this picture, two decades before there were movies to make the ingredients so familiar? Is there a reality behind the romance? Does it have anything to do with the historic experiences of engineers, their training, or the initiation rites that take place in schools? And what about those other engineers Kate mentioned, the ones who chose cities and theaters and pretty girls? Why don't they sound as familiar?

The answer begins with the most obvious element of the image. This welter of notions about engineers bears a strong resemblance to turn-of-the-century nostalgia for the American west. Except for their special knowledge and the nature of their work, these engineers could about as easily be wild west heroes. And not only do they display the same personal qualities, they embody the idea—so prevalent in the country's past—that nature improved is nature beautified. So it comes as no surprise to learn that of the

---

3. Elizabeth Foote, "A Girl of the Engineers," *The Atlantic Monthly* 98 (December, 1906), 387.

fifty or so feature films of the 1920s in which the male lead was an engineer, practically all were set in a frontier locale. Kate's engineers are products as well of the idea that their work is more art than science. It is ingenuity as much as knowledge that brings civilization to the frontier. These men combine in their lives book-learning, hard work, and a sense of craftsmanship. They are closer to tools, materials, and fellow men than to the laws of nature, and, to that degree, are less scientists than the inheritors of an earlier form of engineering described by the term the "mechanic arts."

When people of the nineteenth century spoke of the mechanic arts, it was usually in the context of educational reform. "Useful" knowledge was America's answer to an Old World, aristocratic, self-indulgent learning and to the political system which supported it. That was the message of all those mechanics' institutes of the 1820s and 1830s whose programs aimed at uniting theory and practice, connecting the skills of the artisan with the insights of science. In that combination, useful knowledge would be the means of elevating "the productive classes" to their rightful place in a republic, and for developing the wealth of its great natural resources.[4]

Educational reformers freighted this language with political ideals, but it also expressed a popular conviction that Americans were naturally ingenious. Here was a culturally dynamic image, then, linking national agenda with native disposition, and it pervaded most nineteenth-century experiments in technical education, from mechanics' institutes to polytechnics, to the "People's College" movement of the 1850s and their logical successors, the land grant agricultural and mechanical colleges of the 1862 Morrill Act.

Yet, time after time, in their attempts to establish these new schools, technical educators encountered the strong opposition of high culture advocates, since in America, where intellectuals so long despaired of creating cultural institutions to rival those of Europe, popular interest in practical education seemed certain to undermine any chance of academic equivalence. So, for instance, when Charles W. Eliot returned from abroad in 1865 to take up a teaching post at the newly established Massachusetts Institute of Technology in Boston, his wealthy cousin, the zoologist Theodore Lyman, wrote him, "The only point I don't entirely make up my mind on is the precise place it should occupy in our system of education." What actually troubled Lyman was MIT's practical orientation, not its institutional arrangements, and he went on to argue:

This country is a constant whirl of dollars and railroads and commerce and invention. Every city is capable of a common school, or High School, or School of Arts et Métiers; but it is only at such a place as Cambridge, with its atmosphere of learning

---

4. For more on this subject see Bruce Sinclair, *Philadelphia's Philosopher Mechanics: A History of the Franklin Institute, 1824–1865* (Baltimore: The Johns Hopkins University Press, 1974).

and its learned memories, that we may hope for a *University:* a place where the grander harmonies of the Universe may be studied, whether those harmonies be illustrated by the relations of Languages, or of the Mathematics, or of the Animal or Vegetable or Inorganic Kingdoms. Until we have such a place, we are an imperfect Nation; excessively developed on the side most wanting among Europeans, but stinted on the side where they are excessively developed.[5]

It was exactly that view of America's cultural needs which in the 1830s led the geophysicist Alexander Dallas Bache, as he came to play a leading role in Philadelphia's Franklin Institute, to terminate evening science courses for artisans in favor of programs of original research and publication. For the same reasons, amateurism was vigorously suppressed within the American Association for the Advancement of Science, although the organization had been started in the belief that science was a democracy of ideas.

But the case of the Lawrence Scientific School at Harvard presents the most egregious example of this competition for cultural hegemony. In 1847 the pioneer textile industrialist Abbott Lawrence donated money to Harvard for a school for young men "who intend to enter upon an active life as engineers or chemists or, in general as men of science applying their attainments to practical purposes."[6] And lest there be any doubt about his intentions, Lawrence ended his letter to Harvard's treasurer with the expressed hope that the funds would be used for these purposes "forever." Despite the clarity of his instructions, they were almost immediately subverted and the school named for him became instead a preserve of pure science ideology while its engineering enrollments languished. Indeed, the common opinion around Boston was that MIT would never have come into existence if the Lawrence School had done the job for which it was established. The lesson men like William Barton Rogers, MIT's founder, drew from these experiences was that they could expect as much condescension from scientists as from classicists, and that their educational dreams were far from safe in the hands of the men from the world of elevated culture.

The problem was especially acute at MIT. Not only did Harvard dominate Boston's academic atmosphere, as soon as Charles W. Eliot became the university's president in 1869, he started what became a thirty-year campaign to take over the Institute, as a way of solving the problem of the Lawrence Scientific School. His periodic amalgamation proposals—in 1870, 1878, and 1897—attracted the engineers because their school suffered badly

5. Theodore Lyman to Charles W. Eliot, Boston, 8 March 1868. Eliot Papers, Harvard University Archives.
6. Abbott Lawrence's letter of 7 June 1847 to Harvard treasurer Samuel Eliot, laid before a special meeting of the President and Fellows of Harvard College on the same date, was also reprinted for a dinner of the Lawrence Scientific School Association in 1909. Harvard Archives. Margaret Rossiter's 1966 Harvard undergraduate thesis, "Louis Agassiz and the Lawrence Scientific School," is a good source of information about that institution.

for want of funds, and it was tempting to think of the relief a merger would bring. In the negotiations surrounding each attempt, however, the terms Harvard insisted on made clear its sense of MIT's inferior intellectual status. So even in the school's darkest moments, what sustained Rogers, Robert H. Thurston at Cornell, and other technical educators like them, was the vision of an alternative training for engineering, "a culture and mental discipline," as one of them put it, that would be as intellectually valid as traditional forms of higher education.[7]

This different approach to knowledge, based on a broad study of the natural sciences, also suggested a way of life, modes of thinking and doing that would be different from those of men from the mechanic arts, with their social roots in the working class. Thus, the reformers of technical education shifted it away from craftsmanship and toward science, though by no means were they aiming at the science of high culture, pursued in a disinterested manner for its own sake. Furthermore, in its emphasis on hard work and its sense that the school served the needs of poor young men, MIT still retained elements of that earlier democratic idealism.

Nathaniel Southgate Shaler, dean of the Lawrence Scientific School, totally rejected MIT's brand of higher education. A geologist, deeply persuaded of the virtues of classical culture and fiercely determined to rescue the Lawrence School from its old decay, he wrote a scathing attack on the kind of training offered by independent technical schools. The matter was of national importance, Shaler claimed, because most Americans interested in higher education sought practical instruction. But since the technical schools prepared their students for direct employment at what he slyly called "crafts" or "'trades," they could at best offer only a limited sort of education. That was the heart of Shaler's argument and he claimed that the more explicitly training led to employment, the more intellectually restricted it was. By contrast, he held up the "free atmosphere of academic culture" in the universities, where knowledge was pursued for its own sake. Against a narrow training, guided by what he called "the uncivilized humor" of monetary concerns, Shaler urged the enlarged opportunities of the university, with its "educative companionship" and historically proven methods. Finally, in a swipe at the European origins of the polytechnics, Shaler claimed that the universities, better adapted to the conditions of American life, would prove the "epitome of our culture."[8]

Though he never mentioned it by name, no one in Boston could have doubted which independent technical school Shaler had in mind and MIT's president, Francis A. Walker, immediately wrote a hot reply. If universities

7. John R. Freeman to G.S. Morison, Providence, 21 December 1897. Freeman Papers, MIT Archives.
8. Nathaniel Southgate Shaler, "Relations of Academic and Technical Instruction, *Atlantic Monthly* 72 (1893), 259–268.

were so hospitable to enlarged intellectual enterprise, Walker asked, why had the Lawrence School such an unfortunate history? In any event, the crucial point in the organization of technical education was not institutional affiliation, Walker argued, but prejudice. What really happened, he asked rhetorically, when technical education was "put out to nurse with representatives of classical culture?" To what extent were technical students advantaged by academic atmosphere or by what Shaler had called "educative companionship?" Walker's answers rested upon his strong sense of MIT's own particular academic direction. With a biting image, he spoke of college students "loafing in academic groves" and in a casual, ruminative way "browsing around among the varied foliage and herbage of a great university." And he contrasted that aimlessness, as well as the frivolity of the college life style, with the energy and purpose of technical students. Walker also attacked the social distinctions inherent in Shaler's argument. Why should aspiring engineers prefer a school "where the stained fingers and rough clothes of the laboratory mark them as belonging to a class less distinguished than students of classics or philosophy?" MIT students, he claimed, saw the university as a place of "snobbishness," "where fashions are set in respect to student life, conduct, and dress which they have neither the means nor the inclination to imitate."[9]

Then, though he did not mention Harvard by name, Walker drew some strong comparisons of his own—between indolence and hard work, irresolution and ambition, luxuriousness and simplicity. Or, as he summed up the differences between the two schools, "the Institute is a place not for boys to play, but for men to work."[10] These proved sustaining images for people connected to MIT, and they returned to them for nourishment when proposals for merger were raised again in 1897. But even as its alumni population increased in size and importance, and began to articulate their sense of MIT's special significance, the terms of the debate were suddenly and utterly transformed by a huge bequest to Harvard for the purposes of technical education.

The new benefactor was Gordon McKay, a self-made millionaire inventor and manufacturer of shoe machinery who had increased his wealth through fortunate western mining investments, in which he had the professional advice of his Cambridge neighbor and good friend, the geologist Nathaniel Shaler. McKay always believed that it was his mechanical talents which had brought him financial success, and in the last decade of his life he began to think of ways in which his considerable wealth might be used to give young men with a technical bent the formal training he had never enjoyed.

9. Francis A. Walker, "The Technical School and the University," *Atlantic Monthly* 72 (1893), 390–394.
10. Henry G. Pearson, *Richard Cockburn Maclaurin* (New York: Macmillan, 1937), 294.

Naturally, he talked with Shaler about this and just as naturally Dean Shaler saw in that golden treasure the revitalization of the Lawrence Scientific School. The most immediate, unanticipated, and somewhat ironic effect of McKay's bequest, however, was to push MIT's president, Henry Pritchett, to reconsider the virtues of an alliance with Harvard, and in 1904, when he approached Charles W. Eliot with that suggestion, it may have afforded the original inventor of the idea some satisfaction.

In no time, Boston's newspapers were full of the story. The *Record* jumped to the most spectacular conclusion when it reported that Pritchett was to succeed Eliot as president of Harvard, to administer both institutions, and it linked this development to large real estate transactions and the completion of the Charles River Dam, then under construction. The *Transcript* offered its readers a simpler but equally intriguing analysis when it said the whole idea was "one of Mr. Carnegie's schemes."[11]

There were elements of truth in both accounts, but as the weeks went by people started to think about the other implications of such an alliance, and more reasons in favor of some form of union began to surface. One of the most compelling was the threatening educational competition of the state universities of the west, direct descendants of the old mechanic arts tradition. In an area long accustomed to privately supported higher education, the thought of those rapidly expanding public schools "virtually giving away instruction and training," was an awful prospect and it argued powerfully for combination. The only way Harvard and MIT could continue to attract students, despite their higher tuition fees and the greater expense of urban living, according to the *Transcript*, was "by keeping ahead of the procession in methods and results."[12] The costs of that policy, however, clearly demanded the most efficient pooling of resources.

Other facts impressed themselves on the minds of MIT's administrators. Its Boston property had no more space for expansion, though every faculty member felt the need for additional classrooms and better laboratories, not to speak of their wishes for improved equipment and larger salaries. But the annual report of the treasurer for 1903 showed a deficit of $34,000. Worse still, his analysis of the school's financial history over the preceding decade revealed that the gap between expenses and income was steadily widening. On the other hand, Back Bay real estate had risen substantially in value since the Institute was established, and that suggested a move to a less expensive but larger site to take advantage of the increase.

Besides those problems, there were neither dormitory nor athletic facilities for students at the old stand. In MIT's early years, most of its

---

11. *Boston Record*, 27 January 1904; *Boston Transcript*, 23 January 1904.
12. "The Proposed Harvard-Technology 'Merger'," *Technology Review* 6 (April 1904), 184; *Boston Transcript*, January 23, 1904; ibid., 25 January 1904.

students lived at home. And, because of the school's consciously serious style, no provisions for recreation had been made. But by 1904, when the enrollment neared 2,000, a real demand for housing and for the ingredients of campus life had emerged.

By that time, too, some of the school's prominent figures had begun to argue the need for a broader curriculum with wider possibilities for general culture and social exchange. The chief spokesman for this point of view was John Ripley Freeman, a graduate of the class of 1876, former president of the Alumni Association and a successful businessman who also maintained an extensive consulting practice in hydraulic engineering.[13] The great industrial changes of the past several years, and his own widening acquaintance with the leaders of those developments, led Freeman to see the necessity for something more than technical training if MIT's graduates were to fill the top positions in America's corporations. Freeman elaborated his position at length, and the metaphor he used to make his point was particularly evocative. If MIT stayed as it was while Harvard created extensive programs of pure and applied science, the Institute would end up, he said, educating the corporals of industry instead of its captains.

It was not simply the threat of new McKay-financed laboratories that Freeman had in mind; it was Harvard's unquestioned ability to provide its students with a smooth cultural finish. What he sought for MIT, then, was the intellectually stimulating and socially refined atmosphere that colleges provided best. When MIT was founded, close proximity to the city's industrial districts seemed ideal for training men of action. By 1904, Freeman's model was a Cambridge University college, "with its own dining room," leisurely conversations at mealtimes, and perhaps tennis courts in the quadrangle.[14]

Men like James P. Tolman, president of the Samson Cordage Works, were scandalized by that kind of talk. In a neat example of what Thorstein Veblen meant by the instinct of workmanship, Tolman argued that it was not MIT's business to turn out "cultured men, but technical graduates, men whose hope shall not be in pecuniary return, nor cultured tastes, nor the pleasures of fellowship, but hard workers, whose reward shall be in achievement."[15] Clearly, Freeman imagined someone more like a gentleman, comfortable in the boardroom, a man who might also be elected a director of the firm and thus be in the way to membership in those other clubs gentlemen joined.

Freeman's advocacy of broad culture had other foundations, too. In 1901 he had been appointed Chief Engineer to the Committee on the Charles

13. For a brief biographical sketch of Freeman, see Roland Madany, "John Ripley Freeman," *APWA Reporter*, 48 (May 1981), 6.
14. "Address of Mr. Freeman," *Technology Review* 7 (1905), 37.
15. James P. Tolman to Henry S. Pritchett, Boston, 6 March 1905. Freeman Papers.

River Dam, a group empanelled by the Massachusetts Legislature to investigate the possibilities of a dam near the mouth of that river. Ever since Boston's Back Bay had been developed in the middle of the nineteenth century, the possibilities of further riverfront improvement, on both banks, were obvious. But whether the stream could as effectively carry off the sewage that flowed into it, if dammed, and a series of related questions, led the state to appoint an investigating committee, chaired by MIT president Henry Pritchett. Their report, issued in 1903, encouraged the Boston financier Henry Lee Higginson, in concert with Andrew Carnegie and others, to buy up the land adjacent to Soldier's Field, on the Boston side of the river opposite Harvard, in the event MIT might be moved to the site or some other educational use found for it.

Freeman originally had Hamburg's Alster Park in mind as a case where river improvements had produced a water front public park—hydraulic engineering in its most pleasing form. By damming the Charles to create a basin from the river's mouth back upstream for several miles, its ugly mudflats, exposed and smelly at low tide, would always be covered. A dam would also eliminate the river's swift and dangerous tidal currents, making it more usable for boating and other recreational purposes.[16] The leap from that idea to the picture of grassy banks and university grounds sloping down to them, just as at Oxford and Cambridge, captivated Freeman and it became his great crusade to recreate that picturesque setting along the Charles.

None of these arguments persuaded either MIT's faculty or alumni. By May 1904, an aroused group of graduates convened a special meeting to debate the question of merger, as its opponents insisted on calling it. Among them, the most widely voiced reason against union was that in any connection with Harvard, MIT would inevitably lose its independence. That was a shorthand way of expressing the fear that after having struggled for years to gain a place in America's educational establishment, technical training would once again be reduced to subordination in that citadel of liberal culture.

There were also those who worried about the effect on MIT students of too close an association with Harvard undergraduates—partly a reflection of their notion that the college was the resort of "wealthy society students, steeped in gambling, and athletics, and surrounded by an atmosphere of medieval classicism."[17] Hiram Mills, Freeman's old mentor from the Essex Water Power Company in Lawrence, Massachusetts, was strongly opposed to

16. The history of previous attempts to dam the Charles, as well as a lengthy analysis of the problems and Freeman's proposed solutions is found in *Report of the Committee on Charles River Dam* (Boston: Wright & Potter Printing Company, 1903). An amiable and informative history of the river by Max Hall was published in *Harvard Magazine* for July-August, 1984, 35–53.

17. Association of Class Secretaries of the Massachusetts Institute of Technology, Report of Special Meeting, 2 December 1904. MIT Archives.

the site next to Soldier's Field, for example, because Harvard's new football stadium was there. How could the "earnest spirit of work" in MIT students be preserved, he asked, if their classrooms were next to that "pit of excitement?"[18] Charles Jackson had even worse fears. Instead of the attractive array of courses Harvard students could follow as they wished, MIT men were schooled by a pre-determined curriculum "to draw accurate conclusions rapidly," and to "acquire habits of very hard work." So, Jackson wondered, what would happen if the technical students learned "that they must themselves bear with narrowness of education because they have not time or money to spend on liberal education?"[19]

These alliance discussions, pro and con, extended into the spring of 1905, by which time the two institutions had worked out a mutually agreeable plan of cooperation. Its essential details were that MIT would take over all applied science training for both schools, using the income from the Lawrence Scientific School's capital and three-fifths of the McKay bequest, as well as its own funds, to support that effort. In exchange, MIT would relocate on the property across from Soldier's Field on the other side of the bank from Harvard, though retaining its own name and effective control over its own affairs—"home rule," as some called it.

By the time that plan was worked out, though, the opposing positions had hardened considerably. The Association of Class Secretaries reported that its survey showed ninety-five percent of the school's former students believed MIT should maintain its "absolute independence." Scarcely better news came from an administration-conducted poll, which revealed that alumni were opposed to union by a three to one margin—about the same as the faculty vote. However, Pritchett decided that since he had the responsibility for the decision, he also had the power to make it, and he pushed a twenty-three to fifteen vote through the corporation in favor of alliance. Incensed alumni wrote indignantly to each other about "such extraordinary exercise of corporate power," and there was dark talk about having Pritchett's head for it.[20] But in the end, it was a court of law that scuttled the 1905 Harvard/MIT union, not faculty or alumni opposition. One of the conditions of the plan was that MIT realize income from the sale of its Back Bay property, some of which had come to it in the form of a restricted bequest, and the court ruled it could not do that. The court's decision also meant that Henry Pritchett could not successfully continue as MIT president; so when

18. Hiram Mills to Mr. President, and Gentlemen of the Corporation, Lowell, 20 March 1905. Freeman Papers.

19. Charles C. Jackson to John R. Freeman, Boston, 10 December 1904. Freeman Papers.

20. Association of Class Secretaries of the Massachusetts Institute of Technology, Report of Eighth Annual Meeting, 15 November 1904; Circular letter from E.C. Hultman, Secretary of The Technology League, 1 August 1905. Freeman Papers.

he was offered the direction of the Carnegie Foundation later in the year he did not hesitate in accepting it.

Though the turmoil that had convulsed MIT for the better part of a year came to an abrupt, almost anti-climactic end, the factors that had finally made an alliance acceptable to MIT administrators were still there, with all their pressures unrelieved. The proof of their force is that even after George Eastman began his large anonymous donations to the Institute, solving the problem of money for new buildings, the idea of a combined program in applied science continued to appeal to the leading figures of both institutions. As Richard Maclaurin, the new president of MIT, wrote in 1909 to Abbott Lawrence Lowell, the new president of Harvard, a connection between the two seemed "so natural and desireable [sic]."[21]

If Maclaurin, a New Zealand-born, Cambridge University-trained physicist, lured away from Columbia, had a plan to succeed where Henry Pritchett had failed, he did not make it public. Yet from the start, he simultaneously pushed the search for a new location, for more money, and for an accommodation with Harvard. The idea of a new campus had wide support; the 1905 controversy had made that need apparent. And as he looked at various sites in Boston and Cambridge, Maclaurin began to work on Du Pont alumni for the purchase price. He was also successful in getting a million dollar grant from the commonwealth in 1911. Then, in the following year, Eastman, known only as "Mr. Smith" to everyone but Maclaurin, offered two and a half million dollars for buildings on the new site.

In fact, Eastman was in principle disposed to give money to MIT, but in his initial letter to the industrialist Maclaurin framed ahead of time what he thought would prove persuasive arguments. He characterized the Institute as a place of small and local beginnings that had steadily grown to national stature and international reputation. He suggested as well that the school's preeminence was due to the distinctive way it had integrated science into its curriculum. Then, when Maclaurin finally met Eastman, he revealed his hopes for a cooperative effort with Harvard to create an institution without peer in the world.[22]

These were also the ingredients Maclaurin used to fashion a powerful new image for MIT. It was to be "a great *national* school," he emphasized, based on natural science.[23] His demonstrated skill at raising money gave those ideas considerable momentum. In fact, Maclaurin succeeded so well in convincing others of MIT's special destiny that when the negotiations with president Lowell toward an alliance with Harvard were made public in 1912,

---

21. Richard C. Maclaurin to Abbott Lawrence Lowell, 17 June 1909. Lowell Papers, Harvard University Archives.
22. Richard C. Maclaurin to Mr. Eastman, 29 February 1912. MIT Archives.
23. Richard C. Maclaurin, "President's Report, December 1916." MIT Archives.

it finally seemed a union of equals—each in its own sphere—and it is difficult to find in the records even a murmur of dissent.

Maclaurin also recognized the political liability of the Soldier's Field site, tied as it had been to the 1905 plan, and he determined on the Cambridge location the school now occupies. To help with the design and construction of the new buildings, Maclaurin wrote Eastman, the Institute was fortunate to have the assistance of John R. Freeman, "one of the most prominent engineers of the country."[24] What actually happened to Freeman's ideas for the buildings and the culture they were to reflect shows how far Maclaurin had taken the school from its beginnings in the mechanic arts.

John Ripley Freeman was a confident, vigorous, outgoing person, with a capacity for work notable among men who made that trait a virtue. There is also something slightly sporty about him in the photographs one sees, and in his letters of advice to younger colleagues he stressed the importance of being "a good mixer."[25] Freeman had a strong romantic streak in his nature, too. For several years following the 1905 court verdict, he remained committed to the Soldier's Field site, despite its associations, because of his dream of reproducing in buildings of a Tudor-Gothic style the aesthetic and intellectual qualities of Cambridge University along the banks of the Charles. As he wrote to Richard Maclaurin in 1911, after MIT's canny leader had already abandoned the location:

With Harvard thus on one side and Technology on the other side and a broad river parkway between we ultimately would have a setting as fine as the most ambitious architect would desire and one which, as I said in the beginning, would rival that of St. John's College and the others along the river at Cambridge, England.[26]

It tells you something about Freeman that he was arguing this point with a Cambridge graduate; and that it carried no weight at all with Maclaurin says something about him.

Whatever disappointment Freeman had to swallow when Maclaurin settled on the Charles River basin site presumably got smoothed over by the prospect of designing the school's new buildings there. Actually, no one appointed him to be the architect, but there was considerable confusion at first about selecting one and defining his tasks. Besides, Freeman had ideas and, after all, Maclaurin had asked his advice on the best ways to use the property. So, with the wholehearted commitment an older ethic required in

---

24. Richard C. Maclaurin to Mr. Eastman, 29 February 1912.
25. John R. Freeman to H.C. Bradley, Providence, 4 February 1904. As he once said in a 4 December 1904 letter to Charles Whiting Baker, the editor of *Engineering News,* "engineering work is where I find my golf and whist." Freeman Papers.
26. John R. Freeman to Richard C. Maclaurin, Providence, 9 June 1911. Freeman Papers.

engineers, Freeman gave up evenings and week-ends to clear other work away in order that for several months he could devote his entire attention to the project—which he meant to undertake gratuitously. He had already, he said, "a curious fancy for inspecting college buildings . . . at home and abroad," and he grabbed the chance to create the MIT of his dreams.[27] The result of his work was a strange yet revealing mixture of systematic analysis, scientific management, and sentimentality.

Freeman came to the job with his opinions clearly in place. To begin with, he argued that questions of college architecture generally came down to "about one-fifth architecture and four-fifths a problem of industrial engineering."[28] This case particularly called for efficient methods since the gift from that anonymous donor, Mr. Smith, was so plainly spelled out. The two and a half million dollars was to erect the educational structures of MIT's new campus, and, according to faculty estimates, one million square feet of space was needed. So in its simplest terms, the answer was to find construction methods that would yield the best results for two dollars and fifty cents a square foot. There were other prerequisites in Freeman's mind, though, and he listed them, too; the buildings should have "a flood of light," "a flood of fresh air," "an efficiency and avoidance of lost motion," and should be organized on an effective analysis of "the psychology of student life."[29] Obviously, here was a formula for maximum educational production.

Freeman had studied Morris L. Cooke's report to the Carnegie Foundation on academic efficiency and had talked a good bit with Cooke and his mentor in scientific management, Frederic W. Taylor, about the application of those principles to educational institutions. From his extensive business experience in industrial fire insurance, he also knew about contemporary methods of factory construction. For him, then, the design of MIT's buildings resolved itself into the determination of the optimum organization of interior space and, only secondarily, the matter of exterior appearance.

Freeman's "Study No. 7," the result of months of effort, shows both his head and his heart at work. In the relation of spaces he sought the arrangement, as he repeatedly claimed, "of the best modern factories." And to achieve the greatest economies of fabrication, he urged a modular system of "unit sections," so that "every door casing, every window casing, and every sash, so far as possible, is precisely the same dimensions." But most of all, Freeman argued for the use of reinforced concrete, "pre-eminently the building material of today." In his own mind, any of the aesthetic limitations of that material could be overcome with proper study. What he advocated as an exterior finish was that the concrete be tinted "to give a little of the old

27. John R. Freeman, "Study #7," 12. Freeman Papers.
28. *Ibid.*
29. *Ibid.*, 13.

ivory effect and then . . . the surface tooled with pneumatic hammer to give the texture and play of light and shade."[30]

When the engineers of this era wanted to describe generosity of act or spirit, or the potential of one of their fellows for it, they used the word "big." It meant the opposite of pettiness. Alexander McKim, for example, once described the members of MIT's corporation as "big, broadminded men."[31] That language also implied a model of public behavior, especially in the face of adversity. So when Richard Maclaurin ignored all of Freeman's plans, studies, and drawings—hiring instead John D. Rockefeller's favorite architect, Welles Bosworth—and even left him off the building committee when it was appointed, it was important to Freeman not to be seen as a "sorehead." But in his private correspondence he raged at Maclaurin's surrender to "monumentality," and at the sacrifice of money and efficiency to aesthetics. Bosworth had gotten the job, Freeman suggested to one old friend, because of his "charming manners" and their effect on Mrs. Maclaurin, while his own plans, he wrote to Morris Cooke, were discarded by a "beauty doctor, who never in his life had previously been responsible for building an important structure."[32]

It strikes me that one ought to be able to make something of the contrast between what Freeman wanted and what Maclaurin got—the difference between an engineer's sentimental fondness for the baroque, rendered in reinforced concrete, and the physicist's clean, white classic facade of Indiana limestone. Indeed, Freeman's approach exemplifies some of the qualities of technical culture. His assistants measured the buildings of every engineering school of consequence in North America, many of which Freeman visited, too, not only for specific detail but to search out a kind of muscular style. He had notions for the new buildings, and urged the importance of collegiate life—incorporating among other elements a cloister garden to serve that end. But his design, stripped of its rococo embellishment, was that of an education factory.

Maclaurin's neo-classic building bespoke cool rationality. Its pure and severe style seemed to echo the search for abstract truth, though of a different kind than that distilled in Harvard's old brick buildings. Nathaniel Shaler once argued that since engineering students could not be expected to study the classics, one had "to trust to science to win the ends of culture."[33] He meant high culture, of course, which is what Maclaurin had in mind, too.

30. John R. Freeman to Morris L. Cooke, Providence, 22 October 1914. Freeman Papers.
31. Alex. Rice McKim, Remarks at the Tech Club of Providence, R.I., March 1905. Freeman Papers.
32. John R. Freeman to Hiram Mills, Providence, 4 May 1914; John R. Freeman to Morris L. Cooke, Providence, 22 October 1914. Freeman Papers.
33. Nathaniel Shaler, "Natural Science Training for Engineers," *Engineering Magazine* 9 (September 1895), 1026.

Those who saw the buildings for the first time, at the dedication ceremonies in June 1916, were astonished by their appearance: "long and low and stately above the quiet waters of the Charles," one person said, and then added, "you hadn't expected it was going to be quite so splendid." "That stately shrine across the Charles" was another description. "The white and splendid goal" is how still another observer put it, and that is a good way for us to see the meaning of these structures to MIT people.[34] For them the new buildings were evidence that in the realm of elevated accomplishment engineering had come of age, and they said so.

To one degree or another, all the events of the three-day celebration to mark the Institute's move across the river centered on the transition from old to new, from past to future. An extravagantly decorated boat, built along the lines of a Venetian state barge, carried the school's corporate charter and seal, its archives and faculty, from the Boston side of the river to the Cambridge shore. That pageant and a subsequent theatrical production called the "Masque of Power," were the creations of Ralph Adams Cram, an architect steeped in medievalism whose distinctly anti-modern ideas, ironically enough, proved exactly appropriate.[35] The "Masque of Power" was symbolic in form and the story it told was simple. The world was chaos until mankind brought the great forces of nature under control. But to make those powers the basis of a permanent civilization, mankind needed the aid of Righteousness, as well as that of Will and Wisdom—all allegorical figures in the drama. That simple alliance between engineering and noble human attributes was the essence of the story. The production itself, however, was far more elaborate and involved masses of people, highly stylized dances, costumes, music, lighting effects, and conceits such as a steam curtain. The culmination of the story came when Cram, playing the role of Merlin, brought the elements of nature to the feet of Alma Mater, the spirit of MIT, to be held in trust by her for the welfare of future generations.

It was not the moral tone that impressed people, however, but rather the high aesthetic plane of the production which, like the new buildings, seemed another proof of maturation. To the engineers, the pageants provided

the perfect symbol of what has been happening all these years. The necessity of concentrating on the purely utilitarian has passed; the chrysalis has lived out its time and the butterfly of art has crept out to try its wings.[36]

Another interesting metaphor employed in the ceremonies was that of the struggle from poverty and obscurity to achievement and recognition—

34. *Technology Review* 18 (July 1916), 468, 479.
35. See Jackson Lears, *No Place of Grace: Antimodernism and the Transformation of American Culture* (New York: Pantheon Books, 1981), for information on Cram.
36. *Technology Review* 18 (July 1916), 466.

the progress of the boy with shabby coat and jacket who knew his mind was as good as "the boy with better clothes and a more picturesque leisure."[37] Older men, especially those who had vigorously opposed the 1905 union with Harvard, connected that image with the *art* of engineering, the solid kind of achievement that levelled the waters of the Charles River basin, that produced the civilizing essentials which necessarily antedated creations of aesthetic culture; and they would surely have included pure science in that category.

But while these engineers described a world in which their work provided the basis, indeed was the pre-condition for a more exalted intellectual life, MIT's physicist president took the opposite stand. The central message of his December 1916 annual report was that basic research in science "is of the first importance."[38] Simultaneously, Maclaurin swept away the distinctions between applied and pure science and then, in a model that was to become increasingly familiar, made the first dependent upon the second. That relationship was also the one he continued to articulate with such success to Pierre Du Pont and George Eastman. So, not only did engineering find its culture in science, as Shaler had forecast, it found the funds for its educational institutions there, too. Indeed, Maclaurin's reformulation of the Institute's goals and his talent for raising money so energized the school that when in 1917 the court once again ruled against union with Harvard, on the grounds that it violated the terms of McKay's will, the decision hardly changed anything.

To the main characters of this story, however, there was a vital difference between the culture of science and that of engineering, even if the distinctions were more implicit than explicit. Maclaurin, for example, had nothing but scorn for Morris L. Cooke's scientific management analysis of the efficiency of the physics departments of North America's major universities. "It is written," he said in a comparison full of innuendo, "from the point of view of the man who is used to report on the efficiency of a glue factory or soap works."[39] Equally, in their own correspondence with each other, Freeman, Cooke, and Frederic W. Taylor were convinced that a mathematical physicist would never understand the practical realities of industrial management.

There are also in this quarrel the old echoes of class and status, of the difference between study for the love of it and training for the sake of a job. Once again, the men from high culture seemed to have triumphed. At least Freeman felt that way, and he sarcastically called Maclaurin's successful courting of wealthy industrialists the "dinner jacket" part of the job, by which

37. *Ibid.*, 465.
38. Richard C. Maclaurin, "President's Report, December 1916." MIT Archives.
39. *Science,* new series 33 (January 20, 1911), 101–103.

he meant to imply a manipulative kind of task that engineers were not well suited to perform.[40] Such a remark contrasts oddly with his conviction that MIT students needed to acquire the social skills a collegiate environment taught, yet it is all of a piece with that 1916 dedication ceremonies rhetoric about bright but poor boys in shabby clothes. It is also consistent with the notion that engineering was a democracy of the intellect and that its practitioners honored simple, republican virtues.[41] Indeed, this self-image is not very different from that characterization of Kate's in "A Girl of the Engineers"; it conjures up a group of straight-talking, hard-working men who dealt in facts, not in appearances or tricky shadings of language.

The marvellous irony of this view of engineering culture is that it was articulated by men like John R. Freeman, president of the country's largest factory fire insurance corporation, and that one of his chief concerns was to protect MIT's tuition fees and enrollments from the competition of publicly-supported western universities. It is no less paradoxical that this imagery was so forcibly brought into play just as the work and terms of employment for engineers were being altered by the emergence of large, powerful corporations and science-based industries, by urbanization, and by substantial social change. In fact, Freeman knew something of those dislocations, which is why he had advocated reform of the engineering curriculum. But his nostalgic prescription of cloister gardens contrasts sharply with the views of MIT's men of science. A.A. Noyes, for example, head of the physical chemistry laboratory, argued—in the tradition of nineteenth-century science administrators like Alexander Dallas Bache who had also seen in technology the avenue to their ambitions—that instead of the old slogan "knowledge is power," the school's new motto should be "power is the test of real knowledge."[42]

In this new MIT, there was less and less room for those aspects of engineering that once connected it to the mechanic arts and to democratic ideology. As at Harvard, the demand for admission led to restricted enrollment, just as the emphasis on science undermined notions of native ingenuity. And the school's graduates were, less and less, those idealized representatives of American manhood who in an earlier day had carried the country's dreams to faraway places. Instead, like the other engineers in Kate's story, they chose the cities, "near the crowds and the theatres and the girls." The men she admired for their need of "movement and freedom and fresh air" survived only as fictional characters, and their emergence into early twentieth-

40. John R. Freeman to Frederick W. Taylor, 6 November 1914. Freeman Papers.
41. In their "Report of a Special Meeting" on 2 December 1904, the Association of Class Secretaries claimed, for instance, that "the Technology spirit is one of strong and wholesome democracy." MIT Archives. See also the "Address of Mr. Freeman," *Technology Review* 7 (1905), 37.
42. A.A. Noyes, "The Ideals of the Institute," *Technology Review*, 7 (April 1905), 150.

century romantic literature is another way to mark the passage of the profession's pioneer stage.

It is interesting to think that engineers appear in modern popular culture only after, as a profession, they had largely discarded the attitudes and outlook of the mechanic arts, which in previous times so directly connected them to major themes in the American experience. The fictional characters in those movies and novels of the first two decades of the century did not, of course, reflect a historic condition of engineering; theirs was the nostalgic appeal of a mythic American past, in which simple manly virtues and a wealth of natural resources were central elements in the national drama.

That was a powerful image and it infused the rhetoric of MIT's 1916 dedicatory celebration. Thus, even in the midst of Maclaurin's new buildings, and the new ethos of engineering they implied, the profession's heroic past was ceremonially associated with art, not science. Inevitably, that formulation conveys a sense of loss, but it is a measure of the difficulty engineers have had in defining satisfactory symbols for themselves that they continue to employ it. So, according to Samuel Florman (and with an example reminiscent of John Ripley Freeman's fondness for Tudor Gothic), it was the art of engineering which created the glories of Durham cathedral, but engineering as science that is responsible for the freeways of Los Angeles. Similarly, in contemporary science fiction, a literary form substantially produced and read by engineers, it is the most stereotypical of Robert Heinlein's rugged, individualistic, characters who argue that "engineering is an art, not science."[43]

43. Samuel C. Florman, *The Existential Pleasures of Engineering* (New York: St. Martin's Press, 1978), 60. The central figure in Heinlein's *Door into Summer,* for instance, is a tough-minded, good-hearted, independent, and brilliantly inventive engineer who not only embodies the personal qualities technical men today like to think are characteristic of the profession, but also reflects many of the personality traits with which engineers of early twentieth-century popular culture were endowed.

"What the Senate Is to the American Commonwealth": A National Academy of Engineers

Carroll W. Pursell

Within weeks of the American entry into World War I, the journal *Engineering News-Record* declared "The individual engineer stands alone today, as he has for years. But today he feels this condition, and knows it to be bad, a weakness and a symptom of disease. That is the new spirit, a semiconscious, half-formed demand for professional unity. . . . To make out of the country's great army of engineers an engineering profession," it concluded, "is the work that must be done in the immediate future."[1] This sentiment was very much in the air in 1917, and for some of the nation's most distinguished engineers, the answer lay in the revival of a decades-old dream, the establishment of an American Academy of Engineers. The suggestion had been around at least since the 1880s, but new conditions pumped new life into it. The growing number of engineers, spawned by a galloping industrialization, was an underlying cause, made sharper and more specific by the apparent need for mobilization of the nation's engineering resources in the service of war. Within a few months, however, the movement was dead again, killed partly by the lack of that very engineering unity it was designed to create, and perhaps in part by the aggressive organizational success of the country's scientists. Not until the 1960s was the notion successfully revived, but even then under circumstances that made it less than satisfactory.

Because of the urgent demands of an industrializing nation, the number of engineers in the United States had risen from an estimated 8,000 in 1880 to approximately 136,000 in 1919.[2] The American Society of Civil Engineers, first established in 1852 and reorganized in 1867, claimed to be the legitimate voice for this body, but a host of other organizations, based on geography or technical specialization, had risen up to give the lie to that

---

1. *Engineering News-Record* 78 (May 10, 1917): 322.
2. Esther Lucile Brown, *The Professional Engineer* (New York: The Russell Sage Foundation, 1936), 60.

pretension. The other so-called "founder" societies—the American Institute of Mining Engineers (1871), the American Society of Mechanical Engineers (1880), and the American Institute of Electrical Engineers (1884)—had especially become, by 1917, the centers of extensive and influential constituencies. It was, as one scholar has written, "a golden age of prestige for the engineer."[3] Ironically, it was also a period when engineers felt they were very much unappreciated. As Charles Rosenberg has pointed out, it was a time of "the changing nature of organized knowledge and the context in which it was elaborated, transmitted, and used."[4]

The urge to give a unified and authoritative voice to the disparate body of American engineering was expressed as early as 1886 by the mechanical engineer, William Kent. The protégé of Robert H. Thurston, his dream, as he expressed it, was for an American Academy of Engineering, made up of perhaps three to four hundred engineers, "an aristocracy based on intellect and achievement," which would be "powerful in influence for good." Chosen by peers and representing all branches of the profession, the members of the Academy would have "a large fire-proof building in New York City" with meeting rooms, a museum with models and photographs of engineering works, laboratories for research, and a college for teaching "the engineers of the future." The Academy would hold annual meetings, publish proceedings, and conduct research on matters touching upon "the public health or safety" as called upon by the national or state governments. Finally, he carefully spelled out the manner for choosing the membership, beginning with the "presidents and past presidents of all the engineering societies in the United States." It was, as he said, a scheme that "has been growing in my mind for at least two years, getting larger and larger."[5] Addressing a meeting of the American Association for the Advancement of Science, Kent could hardly have been unaware of the way in which his academy paralleled the National Academy of Sciences, then twenty-three years old. He was probably not aware, however, of the way in which this longing for the coupling of research, service, and recognition echoed calls for agencies such as a national university going back to the very founding of the Republic.[6]

Kent's proposal lay on the table for two decades until it was taken up again (or perhaps independently conceived) in 1909 by the St. Louis bridge

3. Monte A. Calvert, "The Search for Engineering Unity: The Professionalization of Special Interest," in *Building the Organizational Society: Essays on Associational Activities in Modern America*, ed. Jerry Israel (New York: The Free Press, 1972), 49.

4. Charles Rosenberg, "Toward an Ecology of Knowledge: On Discipline, Context, and History," in *The Organization of Knowledge in Modern America, 1860–1920,* ed. Alexandra Oleson and John Voss (Baltimore: The Johns Hopkins University Press, 1979), 440.

5. William Kent, "Proposal for an American Academy of Engineering," *Van Nostrand's Engineering Magazine* 35 (October 1886): 277–280.

6. Bruce Sinclair, *A Centennial History of the American Society of Mechanical Engineers, 1880–1980* (Toronto: University of Toronto Press, 1980), 43.

engineer, J.A.L. Waddell, and the electrical engineer from New York, C.O. Mailloux.[7] In the fall of that year the two put forward a proposal to establish an American Academy of Engineers, and a General Committee of Organization was formed for that purpose. Addressing that group at the Hotel Astor in New York late in 1910, Waddell laid out the grand purpose and goals of the proposed academy. The "prime object," he declared, was "to dignify and exalt the profession of engineer in the broad sense, and to place it upon the highest plane amongst the liberal professions." This object would be accomplished through implementation of a series of specific goals which he identified as: 1) "the establishment of a court of last appeal in all matters relating to the profession"; 2) "the influencing of legislation, both state and national, to promote the development of the profession and to take action in worthy enterprises which involve engineering"; 3) "the choosing of engineers for special services, both public and private"; 4) "the extending of American engineering influence abroad, and especially to the Latin-American republics"; 5) "the inauguration of a code of ethics for engineers in general"; 6) "the exchanging of ideas with engineers of foreign countries"; 7) "increase of compensation for engineers"; 8) "improvement of engineering literature"; 9) "encouraging of original research"; and 10) "establishment of testing apparatus."[8] It was a catalog of desiderata that accurately reflected both the grievances and pretensions of the American engineering profession at the turn of the century.[9] The rest of Waddell's address dealt with a method of choosing the membership of this elite body, a problem which was deemed to be of critical importance for its success.

The proposal was only one manifestation of a lively ferment during these years, as engineers attempted to define their role and status in a rapidly industrializing world. In France, for example, a struggle developed between the state corps of engineers and the École Polytechnique, which had dominated the French profession before 1880, and the army of new industrial engineers which was growing in numbers and prestige. The membership of the Société des Ingénieurs Civils de France grew from 2,000 in 1882 to 6,000 on the eve of the First World War.[10] By 1917 Mailloux, who was in close contact with his French colleagues, was able to report that "the celebrated

---

7. C.O. Mailloux to William C. Redfield, 3 March 1917, file 74901, General Correspondence, Office of the Secretary, Dept. of Commerce Records, Record Group 40, National Archives.

8. J.A.L. Waddell, "Address delivered to the General Committee of Organization . . . ," 10 December 1910, in Box 42, George E. Hale papers, California Institute of Technology.

9. See Edwin T. Layton, Jr., *The Revolt of the Engineers: Social Responsibility and the American Engineering Profession* (Cleveland: Western Reserve University Press, 1971).

10. Terry Shinn, "From 'Corps' to 'Profession': The Emergence and Definition of Industrial Engineering in Modern France," in *The Organization of Science and Technology in France, 1808–1914,* ed. Robert Fox and George Weisz (Cambridge: Cambridge University Press, 1980), 203–205.

Académie des Sciences, one of the four great academies which, together with the Académie Française, constitute the Institut de France, had not proved adequate and satisfactory as a means of providing for the further development and utilization of the activities of the engineering profession, and that the necessity for the creation of a French Academy of Engineers was now well recognized."[11]

The National Academy of Sciences in the United States did not seem to hold out any more hope for the recognition of engineers than did its French counterpart. Although a handful of engineers had been among the original members of the Academy in 1863, their numbers had dwindled over the years. In 1872 the Academy, acknowledging the growing body of engineers and inventors, had elected the bridge engineer and entrepreneur James B. Eads to membership. But an attempt to elect John Ericsson backfired when he refused the honor. It seems clear that the Academy was testing the boundaries of science with these elections, and had decided to draw back.[12] After this, the Academy became more and more a method of honoring accomplishment in pure science. After one of Waddell's public speeches in favor of an engineering academy, however, he was followed by a speaker who pointed out that when the Academy of Sciences had been asked by Congress to investigate the building of facilities for the fixing of atmospheric nitrogen, it had had to turn to the engineering societies for the expertise to discharge the assignment, "a very strong" illustration, he said, of the need for a separate engineering academy.[13] In the face of increasing activity by engineers to organize for both unity and wartime service, however, the Academy, at its spring meeting in 1916, voted to establish "a section of engineering . . . which shall include men who have made original contributions to the science or art of engineering." Two or three engineers were to be elected each year, and attached to either the section on chemistry or physics, until a sufficient number had accumulated to form their own section. This plan resulted in the election of some of the top engineering leaders of the country and the establishment, in 1919, of a separate section. Significantly, these engineers were more associated with engineering and industrial research than were Waddell, Goethals, and their friends.[14]

In February 1916, Waddell again publicly took up the cause of an engineering academy, in an address before the Engineers' Society of Western Pennsylvania. When, in December of that year, he began to press actively for its establishment, the issue of technological advice to the federal government

11. Mailloux to Redfield, 3 March 1917, NARG 40.
12. A. Hunter Dupree, "The National Academy of Sciences and the American Definition of Science," in Oleson and Voss, *Organization of Knowledge,* 354.
13. Remarks of Mr. McDonald, in J.A.L. Waddell, *Memoirs and Addresses of Two Decades,* ed. Frank W. Skinner (Easton, Pennsylvania: The Mack Printing Company, 1928), 112.
14. *Report of the National Academy of Sciences for the Year 1917* (Washington, 1918), 25; *Report . . . 1919* (Washington, 1920), 32, 168.

was very much on his mind. Writing to Secretary of Commerce William C. Redfield, he emphasized that "possibly, the most important function of all" would be "acting, when called upon, as an advisory board for Congress, the Cabinet, the State Governments, etc. in matters involving engineering." In so doing, he concluded, "it could be of immense service to the Nation."[15] His vision of service fitted in nicely with the notion of research which, while not so fully developed by Waddell or Kent before him, also dated back to the earliest calls for an academy. It was a combination that easily lent itself to the needs of war, since that ancient art was one of the last to be industrialized.[16]

The charter for the National Academy of Sciences had been adroitly rushed through a midnight Congress in 1863 as an undebated rider to an appropriations bill.[17] Waddell, perhaps unaware of this useful precedent, sought instead to win the assent of Congress through the utility of his plan, the purity of his motives, and the timely patronage of Gen. George W. Goethals, lately the hero of the Panama Canal. His purpose in writing Redfield was to induce him to use his good offices to convince the General to take the lead in selecting the membership of the planned academy. His friend Samuel W. Stratton, Director of the National Bureau of Standards under Redfield, had already broken the ice with the Secretary.[18]

As Waddell conceived it, the American Academy of Engineers was to be an avowedly elitist group, made up of the cream of the senior engineers in the country. He appears to have been most worried about the problem of choosing this membership in such a way as to win legitimacy for the new organization and to forestall the perception of a self-perpetuating oligarchy. According to his plan, Goethals, who surely was above criticism, would select ten initial organizers (including himself); they in turn would select forty more, and these fifty would become the charter members of the academy.[19] What appeared at first glance to be an exercise in elitism was in fact, as Dupree has shown with the National Academy of Sciences, a necessary drawing of boundaries around a profession that was rapidly changing in important ways.

At the same time, Waddell urged his friends to write letters of support to Redfield. Such luminaries as Carl Hering, the Philadelphia consulting electrical engineer, William G. Raymond, Dean of the College of Applied Science at the State University of Iowa, Dugald Jackson of MIT, and Charles Rand, President of the United Engineering Society, did so. On 3 January

15. Waddell to Redfield, 22 December 1916, NARG 40.
16. Compare Gilbert F. Whittemore, Jr., "World War I, Poison Gas Research, and the Ideals of American Chemists," *Social Studies of Science* 5 (May 1975): 135–163.
17. A. Hunter Dupree, "The Founding of the National Academy of Sciences—A Reinterpretation," *Proceedings of the American Philosophical Society* 101 (October 1957): 434–440.
18. Waddell to Redfield, 22 December 1916, NARG 40.
19. *Ibid.*

1917, Redfield wrote to Goethals, sending along a copy of Waddell's Pittsburgh speech and outlining the proposed method of selection. Tactfully, he also suggested that five of the nine initial members be past presidents of the four founder societies and of the Society for the Promotion of Engineering Education.[20]

Before the end of the month Goethals had agreed to take on the task, and proceeded to pick his names. He left himself off the list but added Stratton, who immediately requested that he be left off because of his previous role in enticing Goethals to serve. The exaggerated modesty of both men underscored Waddell's concern over the appearance of self-serving in the selection process.[21] The list, as made up by Goethals, included besides Stratton, Waddell and Mailloux (both of whom were acknowledged as the leaders of the 1909-1910 effort), the chemical engineer C.F. Chandler, the mechanical engineer W.F.M. Goss, Carl Hering, the civil engineer Clemens Herschel, Charles F. Rand, A.N. Talbot, a past president of the Society for the Promotion of Engineering Education, and Stevenson Taylor, past president of the Society of Naval Engineers and Architects.[22] At this point Redfield wrote Waddell that he considered his "pleasant duties" ended, and left it to the latter to call the small group together.[23]

Moving quickly, six of the group met at the Engineers' Club in New York on March first and again on the second. After first electing Goethals as an eleventh member of their group, they discussed the proper mix of engineering specialties among the incorporators, then adjourned until the ninth.[24] Each of the original members set about drawing up lists of prospective members, and the journal *Engineering News* announced publicly that the process had begun.[25] At the meeting on 9 March, the proper representation of engineering disciplines was agreed upon (civil and military being the largest), a working committee met on 24 March, and set the thirty-first as the day to pick the final thirty-nine incorporators.[26] Of the final list of fifty agreed upon that day, only two, Michael I. Pupin and Elihu Thomson, were members of the National Academy of Sciences. This fact underscored the argument that engineers needed their own academy.[27]

When on 2 April 1917, the war which had been so long anticipated

20. Redfield to Goethals, 3 January 1917, NARG 40.
21. Redfield to Goethals, 30 January 1917, NARG 40.
22. Goethals to Redfield, 15 February 1917, NARG 40.
23. Redfield to Waddell, 27 February 1917, NARG 40.
24. Minutes of Meeting of 2 March 1917, in Subject File, George W. Goethals papers, Library of Congress.
25. Lists are preserved in the Goethals papers, L.C.; *Engineering News* 77 (8 March 1917): 413.
26. Minutes of these meetings are also in the Goethals papers, L.C.
27. A printed list, dated 31 March, is in the General Correspondence, Office of the Secretary, NARG 40.

finally came, the timing proved fatal for efforts to establish the academy. On 26 April, the Hon. Richard Wayne Parker, Member of Congress from New Jersey and a friend of Rand, warned that "the House will take up none but war measures this session" and recommended that "it would be well that the Act be so drawn as to make its war purposes plain & be fathered by the chairman of the committee that deals with war necessities."[28] Later that same day he wrote again, saying that he had met with Rep. William Charles Adamson, Democrat from Georgia and chair of the House Commerce Committee, and that the latter had called Redfield "visionary," adding that incorporation was unnecessary, but "if necessary possible under general D.C. & State laws & that he won't have anything to do with it."[29]

Undeterred, Charles F. Rand, secretary of the group, sent Goethals copies of the proposed bill and accompanying petition, dated 3 May, as well as information on the incorporation of the National Academy of Sciences. The point of the latter, he explained, was to advise the appropriate congressional committees that "what we are seeking is a duplication of what has already been done by Congress."[30] The bill was rewritten by the solicitor of the Department of Commerce, and Senator James W. Wadsworth agreed to introduce it into the Senate. He warned, however, that the upper house, too, had a tacit agreement to treat only war measures in its extra session.[31]

Taking his cue, Waddell asked Goethals for letters of introduction to members of Congress, emphasizing that "the main point being that this bill is essentially a war measure, and that the Academy being composed of the leading men from all branches of engineering would be of exceedingly great service to the Government at this time."[32] Redfield warned Waddell that the proposal was "not a war measure, in the sense that the word is understood by Congress. There is so much of immediate military, naval, and financial importance pressing with Congress now," he added, "that I think it would be a serious mistake to urge the Academy Bill at this particular juncture."[33]

The Senate Committee on the Judiciary recommended passage of a bill of incorporation in July, and printed its slightly amended version of the bill along with a number of endorsements from Stratton, Goethals, and others, and a memorandum from Waddell emphasizing the need for a national charter. He touched again on the oft-repeated arguments, adding a few new ones. "The larger part of the civilized world," he pointed out, "will have to be reconstructed after the war; that such reconstruction is almost exclusively the

28. Parker to Charles F. Rand, 26 April 1917, Subject File, Goethals papers.
29. Parker to Charles F. Rand, 26 April 1917, *ibid*.
30. Rand to Goethals, 4 May 1917, *ibid*.
31. J.A.L. Waddell to Redfield, 14 May 1917, NARG 40; Sen. Wadsworth to Goethals, 23 May 1917, Subject File, Goethals papers.
32. Waddell to Goethals, 21 June 1917, *ibid*.
33. Redfield to Waddell, 22 June 1917, NARG 40.

work of engineers; that the European engineers are being killed off by thousands," leaving the task, presumably, to Americans. He suggested that the Academy, if already in existence, would make "officially a stirring appeal through the press to the youths who are about to enter college" that they should study engineering, and added somewhat awkwardly, that "engineering for the next 10 years, at least, is going to be the most lucrative of all the professions."[34]

The Senate passed an act of incorporation on 4 August, but the House was more reluctant.[35] On the recommendation of the Chief of the U.S. Army Corps of Engineers, to whom the matter had been referred, Secretary of War Newton D. Baker finally endorsed the plan as a war measure on 17 August, but by then it was too late.[36] The measure was never reported out of committee in the House and the matter was dropped.

By this time, the efforts to establish the new academy had circulated among the nation's engineers and the idea now found itself ground between the stones of reaction and reform. The former was embodied in the American Society of Civil Engineers, the organization that not only persisted in its claim to be the rightful representative of all engineers, but that took the lead in resisting any attempts at reform in the profession. In January 1918, its board had come out against the proposed academy, questioning its necessity and asserting that if such a need were present, the proper organizational response would be through the Founder Societies. In reporting this action, the *Engineering News-Record* agreed wholeheartedly. "These certainly are not the days," it insisted, "for the perpetuation of undemocratic and aristocratic ideas. With the world fighting for democracy, the engineers surely are out of tune with the world movement in asking Congress to authorize a self-perpetuating aristocracy." At the same time, it shook its finger at the ASCE board as well. "While the board protests, therefore, against the establishment of a body of super-engineers independent of the national engineering societies, it needs to consider why the rate of increase in juniors in the society has diminished since the year 1917. . . . The profession needs integration," it concluded. "It will not be integrated so long as the struggling young men at the bottom are neglected."[37]

The forces of reform, hinted at by the *Engineering News-Record*, were at this very moment on the rise, particularly in the ranks of the recently formed American Association of Engineers. Destined to grow from 2,300 members in January 1919 to 20,000 in September 1920, the AAE was the voice of those younger engineers who saw the Founder Societies as sacrificing their

---

34. "American Academy of Engineers," *Senate Report* No. 86, 65th Cong., 1st sess., (18 July 1917), 3.
35. J.W. Wadsworth to Goethals, 4 August 1917, Goethals papers.
36. Baker to Redfield, 17 August 1917, NARG 40.
37. *Engineering News-Record* 80 (18 April 1918): 747.

interests to those of their seniors.[38] In March 1918, the board of organization sent a ballot to its members asking for an opinion on the desirability of the proposed Academy. The message with the ballot, however, gently guided the membership by denouncing the Academy as undemocratic and un-American, and as attempting to saddle the profession with an "aristocracy of engineers."[39] "This bill," warned the AAE board, "puts it into the power of a group of men who have attained distinction in the engineering profession, to further glorify themselves and to bring into the charmed circle only the men of their choice. . . . The cloak of patriotism is thrown over the bill . . . ," it charged, "but is so diaphanous that it cannot conceal the ambition which is under it."[40]

Waddell was quick to defend his proposed academy, and by implication, his honor. In May 1918, just two months after the referendum was mailed out, he addressed the Association on the virtues of his plans. He quickly dismissed the charge of self-serving by rehearsing the way in which Goethals had begun the list of incorporators. "In every division and subdivision of engineering," he insisted, "there will be found a few individuals of mature years who are possessed of a deep love for the profession, and who are every [sic] ready to subordinate to its welfare their own personal interests." Such a group of men would be "an aristocracy . . . inconsistent with the principles of democracy" only if "democracy means 'Bolshevikism' . . . otherwise no." In reality, "what the Senate is to the American commonwealth," Waddell concluded, "such would be the American Academy of Engineers to the entire engineering profession of this country." In terms that harked back a century to the ideals of the early republic, he dedicated the Academy to the principles of service, altruism, vision, and deep thinking.[41]

In December 1918, just weeks after the Armistice, Waddell appeared once more before the Engineers' Society of Western Pennsylvania to speak on behalf of the academy. No single statement on its behalf so clearly demonstrates the miscellany of claims marshalled by its backers, an untidy agenda which faithfully reflected the strains within the profession during the Progressive Era. The usual arguments about a court of last appeal, the reform of engineering education, and the advising of government were there, of course, but Waddell ranged even more widely. Regretting that knighthood was not conferred in this country, the Canadian-born engineer observed acidly that "about the only recognized distinction in the United States is the possession of the Almighty Dollar." Worse yet, he charged, "generally the public does not inquire at all closely into the ways and means by which it was

---

38. Edwin Layton, "Frederick Haynes Newell and the Revolt of the Engineers," *Journal of the Midcontinent American Studies Association* 3 (Fall 1962): 22.
39. *Engineering News-Record* 80 (28 March 1918): 631.
40. The text is in J.A.L. Waddell, *Memoirs and Addresses,* 105.
41. *Ibid.,* 106–109.

obtained." Minutes later, however, he was proclaiming that "without capital for investment there would be no engineering, and without engineering there would be very few opportunities for capitalists to invest their money; hence it behooves capitalists and engineers to get together as closely as possible and to work in harmony." By the end of his lengthy speech (it had already "far exceeded the time limit that he set," he apologised), he had endorsed road building, city planning, municipal management, the securing of foreign engineering work, the improvement of engineering literature, and a clearing house for unemployed engineers. The academy ideal was lost, buried under the general need for reform and recognition.[42]

In groping for some analogous organization to hold before the skeptical, Waddell could do no better than name the French Académie des Sciences, and its parent organization, the Institut de France. A decade after this failed to secure his beloved academy, Waddell put it in an even larger context in an address before the American Association for the Advancement of Science. His long-deferred dream, he told his audience, was for no less than an American National Institute, modeled on the Institut de France. Whereas the latter was composed of only five academies, with a total of 325 members and associates and 296 correspondents, his American Institute would have 22 academies with a total of 1,900 members. One of the academies, of course, was to be made up of engineers, like the one then, as he delicately put it, "unavoidably lying dormant."[43]

Again Waddell invoked the virtues of service, altruism, vision, and deep thinking, and again he lamented the reign of the "Almighty Dollar" and the absence of knighthood in America. If the last was too "aristocratic for the American, it ought not to object to a few of its most intellectual citizens wearing, in the buttonholes of their coats, a small piece of red-white-and-blue twisted cord to indicate that they are members of the most exclusive and distinguished of all American societies—The American National Institute." No country, he charged, "can suffer from the existence of an aristocracy of brains; and unless one believe in the absurd theory that all men are equal, he must concede that such an aristocracy is not only unavoidable but also advantageous." The best minds of other nations, too, could become honorary members, and these would, "by keeping in close touch with the Institute . . . carry American influence into the most remote corners of the earth—to the ultimate benefit of the nations and to the aggrandizement of the U.S.A."[44]

Waddell tried to put the best possible face on the failure of the American Academy of Engineers in 1917, which he blamed in part on the "opposition which developed among the rank and file of technicians through the

42. *Ibid.*, 62–75.
43. *Ibid.*, 1148–1149.
44. *Ibid.*, 1157–1158.

newly-constituted American Association of Engineers at Chicago." But he claimed that in part the failure was also a consequence of the establishment in that same year of the Engineering Council, which, he claimed, "the efforts of the Academy's projectors forced the four national engineering societies into founding for the purpose of assuming the many long-neglected professional duties that the Academy aimed to undertake."[45] If Waddell perhaps claimed too much credit for himself and his friends, the Engineering Council could claim little toward providing the unity so many in the profession longed for. Its own Secretary, Alfred D. Flinn, admitted at the end of the war that "we have been muddling along in an unengineering fashion," and, as a result, "the organization of engineers in America is chaotic, complex and illogical."[46]

And so it remained. The eventual establishment of the National Academy of Engineers, in 1964, was as much a sign of weakness as of strength for the profession. Set up under the original charter of the National Academy of Sciences, the Academy of Engineers still lacked that separate and official recognition that Waddell and his colleagues believed could only come through the explicit will of Congress. Even today, when technology appears to be challenging science for federal subsidies, engineering is often defined as a profession rather than a field of learning, and therefore denied the recognition and subsidies bestowed so lavishly on science. Rosenberg has written that "in studying the institutionalization of knowledge in late nineteenth-century America, perhaps the most useful distinction to be made is that between the professions and the learned disciplines." The most important difference between the two, he suggests, is their "relationship to the society that supports them." The learned disciplines receive support indirectly, primarily through universities, while the professions "have related far more directly to their supporting social substrate. . . ."[47]

The frontispiece of the 1928 edition of Waddell's memoirs shows him in cutaway coat, with white starched shirt, white bowtie, white ringletted hair, and a chest resplendent with the honors bestowed upon him by foreign nations. It was the kind of symbolic recognition that he passionately believed to be the just desserts of the accomplished engineer, but which his own nation seemed perversely to withhold. Speaking of the United States Senators at the turn of the century, William Allen White claimed that they represented "more than a state, more even than a region." They "represented principalities and powers in business."[48] By analogy, Waddell's elite of engineering talent had to settle for the same.

    45. *Ibid.*, 1150.
    46. A.D. Flinn, "Efforts to Consolidate the Engineering Profession," *Engineering News-Record* 82 (9 January 1919): 81.
    47. Rosenberg, "Toward an Ecology of Knowledge," 443.
    48. Quoted in Matthew Josephson, *The Politicos, 1865–1896* (New York, 1938), 444.

Materials and Metaphors: Plastics in American Culture

Jeffrey L. Meikle

Plastics are not materials that many people spend much time thinking about. Even though we constantly use objects made of plastics, we are hardly aware of them. From garbage bags to computer housings, from disposable razors to automobile interiors, we take them for granted. Still, there is something unsettling about plastics. In the opening scene of *The Graduate,* that highly popular film from 1967, a middle-aged businessman mystifies young Dustin Hoffman by telling him, "I just want to say one word to you. Just one word. . . . Plastics."[1] Nervous laughter filled the theaters nearly twenty years ago, and the memory now evokes nervous chuckles, but few of us could say precisely why. We no longer even use the word "plastic" to mean fake or phony, as some of us once did, but it retains the power of making us ill at ease.

A sense of the ubiquity of plastic arises from a more recent expression of popular culture, Mark Helprin's bestselling novel *Winter's Tale* (1983). One of the growing number of fictional works that confront the approaching millennium, *Winter's Tale* portrays the experiences of Peter Lake, who plunges from the Brooklyn Bridge at the beginning of this century and magically awakens in a Manhattan hospital near the year 2000. The first thing he sees is "a plastic band around his wrist." We are reminded that "never before had he seen plastic." As he becomes more alert (and consequently more disoriented), he notices that "the way things were shaped, and the materials of which they were made, seemed almost otherwordly," and that "everything seemed to have grown smooth, to have lost its texture." In the new world that Peter Lake has entered, there is no iron or wood to be found.[2] He has landed, in other words, right in the middle of the "Plastics Age," which began in 1979, according to an industry spokesman, when the annual

---

1. As quoted by Sylvia Katz, *Plastics: Designs and Materials* (London: Studio Vista, 1978), 5.
2. Mark Helprin, *Winter's Tale* (New York: Pocket Books, 1984), 373, 378

volume of plastics produced in the United States exceeded that of steel.[3] But once novelist Helprin has registered his time traveler's first response to the unfamiliar materials in his new enviroment, plastic is as much taken for granted by Peter Lake as it is by the rest of us. Noteworthy at first glance, it soon becomes invisible.

Despite the everyday "invisibility" of plastic, common sense suggests that it has had a profound impact on life and thought in the twentieth century. As far back as 1927, publicists for the industry proclaimed a "Plastic Age" equal in historical significance to earlier ages of bronze, iron, gunpowder, or steel.[4] Williams Haynes, an economist and historian of the chemical industry, declared several months into World War II that plastics and other synthetic materials would have "more effect on the lives of our great-grandchildren than Hitler or Mussolini." Adopting a moderate technological determinism, Haynes maintained that "new materials can compel the course of history as greatly as any man."[5]

Such a statement is difficult to prove, especially if one is looking for more than tangible economic impact. Plastics themselves, by their very nature, complicate the problem of determining their cultural significance. Able to assume virtually any shape, texture, hardness, density, degree of resilience, or color, the myriad varieties are united only by a word, "plastic," that has defied most attempts to promote one trade name over another. What do we mean when we talk about plastic? Is it a material truly capable of shaping the course of history, or is it merely an amorphous substance, receptive to virtually any psychological or cultural projection?

More than twenty years ago, Marshall McLuhan faced a similar question when exploring the cultural impact of the electronic communications media. He maintained that a new medium tended to upset the delicate ratio among the human senses by elevating one at the expense of the others and by numbing any awareness of the imbalance. Although McLuhan's theory remained unproven and probably unprovable, it is possible that changes in the ratio of the *materials* of the everyday environment affect consciousness and, consequently, the forms of cultural expression. During the twentieth century, the ratio of plastics and other synthetics to wood, metals, cotton, wool, and other traditional materials has steadily increased. Our unconscious acceptance of this fact corresponds to McLuhan's concept of numbness. To dramatize society's response to new communications media, McLuhan

---

3. Sidney Gross, "Plastics Age Arrives Early," *Modern Plastics* 57 (February 1980): 45.
4. "Editorial Impressions: We Believe in Dreams!" *Plastics* 3 (February 1927).
5. As quoted in "Industry Prepares for War Role," unpaginated "special insert" in *Modern Plastics* 19 (May 1942), reporting on a convention of the Society of Plastics Industry. See Herman F. Mark, "Giant Molecules," *Scientific American* 197 (September 1957): 89, for similar comments from a prominent chemist.

announced that "we become what we behold."⁶ Considering our response to new materials like plastics, one might suggest that "we become what we *mold*"—or make, or use.

To accept this analogy from McLuhan brings us no closer to an understanding of how plastics have shaped contemporary culture. However, evidence that artifacts do affect us is provided in a study by two social scientists, *The Meaning of Things: Domestic Symbols and the Self* (1981). Mihaly Csikszentmihalyi and Eugene Rochberg-Halton analyze the household possessions of a cross-section of Chicagoans. The domestic environment, they maintain, makes up "an ecology of signs that reflects as well as *shapes* the pattern of the owner's self."⁷ Basing their conclusions on questionnaires, interviews, and psychological case studies, the authors demonstrate that personal possessions embody goals, define behavior, and become extensions of their owners' personalities. However, *The Meaning of Things* discusses only objects singled out by their owners as prized possessions. Its methodology suggests no techniques for assessing the significance of the myriad other everyday objects—many made of plastics—whose meanings probably cannot be articulated by their owners.

How, then, to arrive at any understanding of the relationship of plastics and culture? McLuhan's vision suggests by analogy a bold, attractive hypothesis, but it cannot be proven. The Chicago researchers, on the other hand, document a relationship between things and ideas, but their method cannot be extended to objects normally beneath conscious awareness—nor to materials from which practically anything can be made. How do *these* materials reflect intentions and shape behavior?

In what follows, I suggest a tentative outline of the relationship between plastics and culture by tracing two historically linked developments: first, ideas about plastics expressed by the industry and its publicists; and second, the accelerating applications of plastics themselves. Ideas about the potential of plastics emerged fully developed early in the twentieth century, at a time when plastics themselves were relatively scarce. Early promoters conceived of them as utopian materials and, perhaps because the most visible plastics were of the permanently stable thermosetting variety, conceived of plastic objects as symbols of a static, eternally perfect future society—an antidote to the chaos of the 1930s and 1940s. Then, as chemists created new varieties of plastics and manufacturers applied them in ever greater volume, a shift in emphasis occurred. The new postwar materials were of the thermoplastic variety, infinitely capable of being melted and reshaped. Their lack of stability

6. Marshall McLuhan, *Understanding Media: The Extensions of Man* (New York: Signet, 1966; originally published 1964), 33.
7. Mihaly Csikszentmihalyi and Eugene Rochberg-Halton, *The Meaning of Things: Domestic Symbols and the Self* (New York: Cambridge University Press, 1981), ix, 17.

and their dramatic proliferation mirrored the expansive postwar economy and encouraged notions of disposability and impermanence, with regard both to the plastics and to society. The utopian vision of the postwar period emphasized not stability but endless change and transformation. As we finally entered the Plastics Age, with synthetics all around us, we began to describe our culture itself with terms and metaphors formerly used to characterize plastics. Ignoring the materials, we projected their values onto our institutions and, in fact, into our lives.[8]

Such a prospect was not envisioned in 1909 by Leo Baekeland, a Belgian immigrant and industrial chemist, when he announced his synthesis of Bakelite, a substitute for shellac made from phenol and formaldehyde.[9] Although he intended Bakelite for industrial applications, he admitted that it might find limited use in such novelty items as buttons, where it would replace celluloid, an early imitative replacement for ivory, horn, and other natural materials. For the next two decades, Bakelite expanded in use as a hidden electrical insulating component of automobiles and radios. But almost from the beginning manufacturers were using Bakelite, the first truly synthetic plastic, in pipe stems, costume jewelry, phonograph records, billiard balls, umbrella handles, and other peripheral consumer products.

A skillfull entrepreneur, Baekeland promoted his discovery as a unique new material of marvellous scope rather than as an imitative substitute. Many prospective business customers learned of the plastic through an elegant, evocative little book, *The Story of Bakelite,* written by John Kimberly Mumford. With 25,000 copies distributed by 1927, three years after its publication, the volume represented a considerable promotional success.[10] Mumford described Bakelite as "a wonder-stuff" that had emerged miraculously from "the world's waste heaps"—the coal that had been cooking in the "silent laboratory of the earth" since the dawn of time. From practically nothing had come infinite riches. Above all else, in Mumford's opinion, Bakelite exhibited "a Protean adaptability." It could fill virtually any need and assume any desired shape. But once heat and pressure had given it the strength of its final

---

8. For discussion of this projection, see the author's essay on "The Culture of Plasticity: Observations on Contemporary Cultural Tranformation,"*Amerikastudien/American Studies* 28 (1983): 205–18.

9. Leo Baekeland, "The Synthesis, Constitution, and Uses of Bakelite," *Journal of Industrial and Engineering Chemistry* 1 (March 1909): 156–57. On the economic and cultural significance of celluloid, see Robert Friedel, *Pioneer Plastic: The Making and Selling of Celluloid* (Madison: University of Wisconsin Press, 1983). The only general histories of plastics are M. Kaufman, *The First Century of Plastics* (London: The Plastics Institute, 1963), and J. Harry DuBois, *Plastics History U.S.A.* (Boston: Cahners, 1972), both of which focus primarily on technical developments. Excellent photographs of plastics applied to consumer goods can be found in Andrea DiNoto, *Art Plastic: Designed for Living* (New York: Abbeville Press, 1984) and in Sylvia Katz, *Plastics: Common Objects, Classic Designs* (New York: Harry N. Abrams, 1984).

10. For publication information see "How Bakelite Uncovered New Uses for the Product," *Sales Management* 13 (24 December 1927): 1118, 1160.

form, it would "continue to be 'Bakelite' till kingdom come."[11] A protean material, Bakelite was cheap, inexhaustible, versatile, durable, and somehow exotic, even though it promised to usher in an age of material abundance and democratic luxury. Unlike celluloid, which thrived on the imitation of so-called natural materials, Bakelite was frankly artificial: a product of chemical genius, possessing unique qualities of strength and beauty. Virtually every publicist or journalist who discussed plastics in the years before World War II did so in similar utopian terms.[12]

The Bakelite Corporation signaled its own utopian aspirations by describing its product as "the material of a thousand uses" and by adopting the mathematical symbol for infinity as a trademark. For many applications, Bakelite seemed "a better material than any which Nature unaided has provided."[13] During the 1920s, the company attracted commercial business by offering technical assistance to manufacturers and by exhibiting sample products made from the new plastic in a score of cities. Publicity also reached the general public through films distributed to schools and service clubs. A typical general-circulation magazine advertisement depicted a medieval alchemist in his laboratory and implied that Bakelite was created by a chemical process as miraculous, as lucrative, as the transmutation of lead into gold.[14] By the end of the decade, the message had been received. Paul T. Frankl, a designer whose machine-age furniture and interiors represented the height of fashion, declared in 1930 that Bakelite and other "materia nova" offered proof that "industrial chemistry today rivals alchemy." Rather than using plastics to imitate traditional materials, it was necessary "to create the grammar of these new materials" so they would "speak in the vernacular of the twentieth century."[15]

Two years later, an advertising executive counseled the industry to make the general public "plastics conscious" by "avoiding imitations and taking the initiative in styling."[16] Always a step ahead of competitors, who by then were taking advantage of lapsed patent rights, the Bakelite Corporation

11. John Kimberly Mumford, *The Story of Bakelite* (New York: Robert L. Stillson, 1924), 7, 20, 10, 46, 51.
12. For more complete coverage of plastics utopianism, see the author's essay on " 'The Material of a Thousand Uses': Plastics in the World of Tomorrow," in *Past Visions,* ed. Joseph J. Corn, to be published by MIT Press.
13. Allan Brown, "Bakelite—What It Is," *Plastics* 1 (October 1925): 17. See also A.C. Tate, "The Protective Power of a Good Trade Mark," *Plastics* 1 (November 1925): 53.
14. See Charles G. Muller, "Bakelite's Round-Robin Advertising Comes Home to Roost," *Printers' Ink* 131 (7 May 1925): 133–34, 136, 139, and John Allen Murphy, "The Bakelite Caravan—A New Idea in Industrial Selling," *Advertising and Selling* 8 (20 April 1927): 23–24.
15. Paul T. Frankl, *Form and Re-form: A Practical Handbook of Modern Interiors* (New York: Harper, 1930), 163.
16. L.W. Hutchins, "Enlarging the Market for Molded Products," *Plastics and Molded Products* 8 (November 1932), 416.

organized a conference in 1933 to encourage industrial designers to explore the aesthetic possibilities of plastics.[17] As a result, such designers as Raymond Loewy, Norman Bel Geddes, and John Vassos soon appeared in testimonial advertisements in *Modern Plastics* and *Sales Management*. Often hailed as a panacea for business during the Depression, industrial design shared its visionary aura with the plastics industry. Streamlining, a design mode that reflected the public's desire for frictionless motion into a static "world of tomorrow," became the preferred style for molded plastic products. A mold with rounded edges and corners could be machined cheaply, and facilitated the flow of material during the molding process. Wide curves helped limit breakage of completed parts. Form and function thus coalesced in a manner rare in the history of consumer product design.[18] And although radio cases provided the most typical expression of "utopian" design in plastic, by 1935 journalists were predicting that automobiles would soon "have molded plastic bodies, and airplanes molded wings."[19]

At the very end of the decade, introduction of a dramatic new synthetic increased public interest in "miracle materials." Such innovations as vinyl, cellulose acetate, and urea formaldehyde had extended the uses of plastics beyond those of Bakelite, but none attracted as many headlines as nylon, invented by Wallace Hume Carothers and announced by Du Pont in October 1938. Although made from "such common raw materials as coal, water, and air," according to a spokesman, nylon could "be fashioned into filaments as strong as steel, as fine as the spider's web, yet more elastic than any of the common fibers."[20] Intended as a substitute for silk, whose supply was controlled by the Japanese, nylon caught the attention of women who anticipated run-less stockings. The first nylon stockings graced the legs of the college girls who guided visitors through Du Pont's pavilion at the New York World's Fair of 1939, and commercial varieties finally went on sale early in 1940.[21]

17. "Plastics in Pictures," *Plastic Products* 9 (March 1933), between pp. 22–23.

18. For discussions of the connection between plastics mold technology and streamlining, see Franklin E. Brill, "What Shapes for Phenolics," *Modern Plastics* 13 (September 1935): 21; Montgomery Ferar and Carl W. Sundberg, "Three-in-One," *Modern Plastics* 14 (January 1937): 55–56; Raymond P. Calt, "A New Design for Industry," *Atlantic Monthly* 164 (October 1939): 541–42; and John Sasso and Michael A. Brown, Jr., *Plastics in Practice: A Handbook of Product Applications* (New York: McGraw-Hill, 1945), 23–24, 89.

19. Elinor Hillyer, "The Synthetics Become the Real," *Arts and Decoration* 42 (January 1935): 27.

20. Du Pont vice-president Charles M.A. Stine as quoted in "Du Pont Launches Synthetic Silk," *Business Week* (29 October 1938): 18. Oddly enough, some reports emphasized that one ingredient used in producing nylon was cadaverine, a substance produced by decaying corpses but also capable of synthesis from coal tar derivatives. See "Castor Oil, Coal Newest 'Silkworms' for Stockings," *Science News-Letter* 34 (1 October 1938): 211, and "No. 2, 130, 948," *Time* 32 (3 October 1938): 47.

21. "Nylon Hose Go On Sale Wednesday," *Business Week* (11 May 1940): 38.

Much of the publicity for nylon seemed frivolous even then. Some observers realized, however, that Carothers' calculated experiments with polymerization held tremendous promise for the future of plastics. The work at Du Pont suggested that it would soon be possible to "custom-build" or "tailor-make" an infinite number of plastics, each with precisely designed characteristics, and all derived from cheap, inexhaustible materials.[22] In a review of the industry in 1940, *Fortune* summarized the typical view by stating that "these tough, utilitarian, and jewel-like materials" were being "pressed and squeezed and rolled and sawed and drawn and cast and carved into the very image of a new world." But the magazine went beyond the "world-of-tomorrow" vision of reassuring stability and permanence to hint that "only surrealism's derangements" could "capture the limitless horizons, strange juxtapositions, [and] endless products of this new world in process of becoming."[23] This comment indicated more accurately than others the expansive, even explosive future of plastics in American society, as thermosetting varieties eventually yielded to thermoplastics.

As of yet, however, the public had experienced plastics only in limited applications: molded cases for radios and electric shavers, Formica counters in banks and diners, steering wheels and control knobs, picnicware and bathroom tumblers, costume jewelry and commercial premiums, and such imitative uses as hairbrushes and eyeglass frames. Many of these products vanished in the homefront austerity of World War II. But plastics remained in the public eye. Along with "Lucky Strike Green," they went to war. *Life* announced in 1943 that the plastics industry, formerly "the nation's biggest and most clamorous maker of gimcracks," had undergone "metamorphosis into a semi-secret, high-priority war industry."[24] Many wartime developments hardly remained secret, however. Newspaper stories about Plexiglas bomber cockpit covers fueled visions of postwar bubble-domed automobiles. New applications of phenolic-bonded plywood to aircraft and emergency housing suggested a future of mass-produced prefab houses, each with a plywood airplane in the garage.[25] Helmet liners, gas masks, bayonet scabbards, bugles, and other military accessories demonstrated the utility of plastics. A documentary film entitled *This Plastic Age,* released in 1942 by the Society of the Plastics Industry, explained "the use of plastics in the planes, the tanks, the guns and the ships" of the American war effort. With an eye on

---

22. See Joseph L. Nicholson and George R. Leighton, "Plastics Come of Age," *Harper's* 185 (August 1942): 301, and "Test-Tube Marvels of Wartime Promise a New Era in Plastics," *Newsweek* 21 (17 May 1943): 42.

23. "Plastics in 1940," *Fortune* 22 (October 1940): 90, 89.

24. "Plastics: War Makes Gimcrack Industry into Sober Producer of Prime Materials," *Life* 14 (3 May 1943): 65.

25. Howard Hughes' monstrous plywood transport plane, the "Spruce Goose," seems absurd only in retrospect.

homefront morale, the film ended by predicting a postwar future "enriched with plastics—plastics for cars, for civilian planes, for furniture, refrigerators, objects of art, for clothes, for better homes and for better living."[26]

Only a year later, the Society of the Plastics Industry had begun to regret its promotion of a peacetime plastics cornucopia. Late in 1943, the Society formed a postwar planning committee entrusted with the ironic goal of "deglamorization." It was necessary to counteract the perception of plastics as a " 'miracle whip' material with which anything can be done."[27] This quick reversal indicated the bleaker side of wartime plastics. All too often the products available to civilians were shoddy, second-rate substitutes, ersatz goods made from contaminated scrap or types of plastic unsuited to their applications. Kitchen utensils of cellulose acetate melted in boiling water, vinyl combs dissolved in hair cream, laminated mixing bowls splintered apart. A marketing study published soon after V-J Day discovered "prejudices . . . that may take a long time to overcome."[28] Although the war had expanded the utopian dimension of plastics, it had also awakened a sense of their inferiority to "natural" materials. As a result, the industry toned down its rhetoric during the decade after the war, at the very time that plastics began to contribute in a visibly prominent way to the shape of consumer culture.

At the end of the war, few people had any desire to let the world of tomorrow take over their homes. With the return of peace and the hesitant emergence of prosperity, they tended to prefer the reassurance of tradition to the smooth surfaces of the machine age. Plastics invaded the home, but they did so under the guise of other materials. Vinyl floor tiles and upholstery masqueraded as marble and leather. Wood-grained Formica protected table tops and wall surfaces. Rayon draperies and lampshades of cellulose acetate appeared strange only on close examination. In the words of Mary Roche, a *New York Times* columnist on interior decoration, the much-touted "house of plastics" had once seemed "outlandish." Now, however, "you might look right at it without even recognizing it" because designers had become "more interested in blending synthetic products with other materials that do not come out of a test tube."[29] Women's magazines emphasized that plastics did not have to be "shiny, sleek, and a little too strange-looking for the living

---

26. " 'This Plastic Age'—New Motion Picture," *Modern Plastics* 19 (May 1942): 116.

27. SPI executive vice-president William T. Cruse as quoted in "Fight Glamourizing of Plastics Role," *New York Times,* 1 September 1943, p.31.

28. Alfred Auerbach, "I Don't Know," *Modern Plastics* 23 (October 1945): 102. See also " 'Information, Please'," *Modern Plastics* 21 (June 1944): 168; Josephine von Miklos, "By Their Lines Ye Shall Know Them," *Modern Plastics* 22 (November 1944): 119–20; "Propaganda . . . a Threat or a Boost," *Modern Plastics* 22 (March 1945): 93–94, 198; and James J. Pyle, "New Horizons in Plastics," *Science Digest* 18 (August 1945): 85–86.

29. Mary Roche, "New Ideas and Inventions," *New York Times Magazine,* 10 November 1946, p.42

room."[30] No longer streamlined, plastics might just as easily assume "homey, chintzy, and comfortable" forms to become plastics that "don't look it."[31]

In October 1947, *House Beautiful* devoted forty-five pages to illustrating and describing the wonders of plastics. Advising readers to "forget the dream world stuff you've heard," the editors got right down to the business of explaining exactly why these self-effacing materials were so important. There was, they argued, "only one good reason why you, personally, should be interested in plastics." That reason was *"damp-cloth cleaning."*[32] Unscuffable, stain-resistant, and virtually indestructible, plastics thus took their place in the postwar domestic landscape as the materials of what might be called damp-cloth utopianism.

As long as the public was learning to accept plastics in the home, the industry gave up its former "machine-age" campaign to eliminate imitative applications. An occasional holdout resisted this trend toward merging the plastics with traditional materials and styles. In 1949, for example, the Admiral Corporation invested a quarter of a million dollars to manufacture a one-piece Bakelite television cabinet weighing thirty-five pounds, the largest civilian molding yet produced. Commenting on the development, the president of the molding company predicted that manufacturers would "cease trying to make plastic products look like wood and concentrate on making the quality of plastics more attractive."[33] But wooden television cabinets became the standard, replaced by plastic only in the 1960s when high-density foams approximated the appearance of wood.

Few businessmen active in the plastics industry questioned imitative plastics. They reserved their concern for shoddy products that damaged the reputation of all plastics. Most frequently cited were the infamous vinyl shower curtains, of which it was said, "they smell, they sweat, the print comes off and they get brittle."[34] Despite the guilt by association of such setbacks, the industry grew rapidly, expanding 600% from 1939 to 1949. In 1950 there were thirty producers of synthetic resins supplying some 3500 molders, laminators, and fabricators, the majority of whom were independents doing contract work for general manufacturers. Growth brought confusion among molders, manufacturers, and consumers because few could keep track of the many new varieties of plastic, their conflicting qualities, and their limitations.

---

30. "Plastics: A Way to a Better More Carefree Life," *House Beautiful* 89 (October 1947): 141.
31. Christine Holbrook and Walter Adams, "Dogs, Kids, Husbands: How to Furnish a House So They Can't Hurt It," *Better Homes and Gardens* 27 (March 1949): 37–39.
32. "Plastics: A Way to a Better More Carefree Life," p.122. The refrain was endlessly repeated in periodicals of the late 1940s and early 1950s.
33. "Admiral Video in Plastic," *New York Times,* 12 May 1949, p.47.
34. "The Shower Curtain Problem," *Modern Plastics* 24 (July 1947): 5.

In general, however, no one questioned the convenience of plastics in everyday life.[35]

During the 1950s, as plastics proliferated in the ranks of consumer goods, their essential nature altered. Until then, most plastics were "thermosetting"; that is, their formation under heat and pressure was an irreversible process. Bakelite's stability and durability had suggested to early promoters a kind of material immortality—a reputation later extended to other phenol and urea formaldehyde plastics, the materials most commonly used for radio cabinets, dishware, cameras, and other molded products. The forms of many of those products reflected the desire for permanence and stability, embodied in the preeminent Depression style of streamlining. During the 1950s, however, the plastics industry shifted from thermosets to the so-called "thermoplastic" materials, which could be melted and reshaped an infinite number of times after their initial polymerization.[36]

Thermoplastics like polyethylene and polypropylene contributed to a flood of new applications—squeeze bottles, Tupperware, hula hoops and other flexible toys, garbage pails, packaging, and so on, with new products following in geometric progression. The wave of new thermoplastics in turn stimulated a renewed interest in vinyl, an older thermoplastic with parallel uses. More important, it marked a shift in raw material from coal to petroleum. And the most obvious distinction between coal and petroleum—that one is a dense solid, the other a viscous liquid—symbolized the difference between thermosets, eternally permanent in form, and thermoplastics, with their capacity for infinite shape-shifting. During the 1950s, however, an affluent America had become too preoccupied with the ever-changing flow of consumer goods to think of such analogies. The metaphoric implication of the new plastics remained largely unconscious and was perhaps therefore even more compelling than the 1930s image of permanence and perfection—the teardrop-shaped Bakelite housing. By the time the metaphor became conscious, during the late 1960s, thermoplastics had already materially and visually contributed to the explosion of a shapeless, ever-changing, impermanent, ephemeral culture of consumption.

At first, many consumers resisted the thermoplastic revolution. Manufacturers of packaging materials found that the public rejected the idea of plastic "disposables." Paper products seemed more appropriate owing to "the ease with which they may be made to disappear after use."[37] But as plastics continued to fall in price, they became competitive with paper.

35. For a general survey see "A 1950 Guide to the Plastics," *Fortune* 41 (May 1950): 109–118, 120.
36. See "The Polyethylene Gamble," *Fortune* 49 (February 1954): 134–37, 166, 170, 172, 174; and Francis Bello, "The New Breed of Plastics," *Fortune* 56 (November 1957): 172–75, 218, 220, 223–24.
37. "Plastics for Disposables," *Modern Plastics* 33 (April 1956): 5.

Manufacturers faced the problem of educating the public to see plastics as disposable products. An article published in 1957 bemoaned "the disinclination of consumers to accept the fact that such merchandise has been designed to be, and therefore should be, discardable and destroyable."[38] The issue came to a head with tragic irony in 1959, when more than fifty infants and children were suffocated by polyethylene dry-cleaning bags, many of which had been re-used as mattress covers. Although Du Pont blamed "parental carelessness" and the editor of *Modern Plastics* complained that the film had been "made and costed to be disposable," only a year earlier the magazine had promoted the handy bags because they could be "re-used by the customer."[39] To get people to accept plastics for packaging in the first place, manufacturers had emphasized durability and re-use. Now, however, it seemed that a "profitable . . . share of plastics' future" was "in the garbage can." Americans had to be persuaded that not only packaging, but also disposable products, would contribute to the new age of "casual living and increased leisure" they had recently entered.[40] As it turned out, increased production of disposable plastics contributed to a more general awareness of the ephemeral nature of experience itself.

At the same time, although domestic uses of plastics continued to emphasize imitation, a few observers began again to celebrate liberation from tradition through frankly artificial materials. In 1957, polymer chemist Herman F. Mark described the ability of researchers "to devise new materials which nature has neglected to create." Rather than enduring the limitations of natural materials, the human race could enjoy "an almost limitless number of combinations," each "tailored" to satisfy a specific need.[41] Thinking along the same lines, a journalist concluded in 1961 that plastics, "the world's first 'common denominator' materials," would yield a higher standard of living for all areas of the world. Capable of being made from coal, petroleum, or agricultural waste, depending on local supply, the plastics possessed "a wider range of available properties than in all other materials combined" and were "suitable for use in any earthly clime, or even in outer space."[42] Similar humanitarian, even visionary concerns dominated discussions of plastics in architecture. Enthusiasts hoped that synthetics would make cheap, adaptable, durable housing available to the masses. Most prototype dwellings of plastic

---

38. "Plastics in Disposables and Expendables." *Modern Plastics* 34 (April 1957): 93.

39. Morris Kaplan, "Industry Warns on Plastic Bags," *New York Times,* 4 June 1959, p.33; Hiram McCann, "Hazards in Film Misuse Must Be Taught Parents," *Modern Plastics* 36 (June 1959): 262; and "Dry Cleaning: Big New Market for Film," *Modern Plastics* 35 (April 1958): 106–107.

40. "Plastics in Disposables and Expendables," 206.

41. Herman F. Mark, "Giant Molecules," 89 (see note 5 above). See also a breathless survey directly inspired by Mark's article: Francis Bello, "The New Breed of Plastics," previously cited in note 36.

42. "The Wide World of Plastics," *Modern Plastics* 38 (May 1961): 129.

managed to look as artificial or nontraditional as possible. In addition, following the general plastics trend, successive designs became less stable, increasingly temporary or ephemeral.

As early as 1954, architecture critic Douglas Haskell articulated the vision behind virtually all experimental applications of plastics to human shelter. Plastic buildings, he maintained, would be "all 'skin'." Beyond that they might be "as thin as egg shells, as ribbed as leaves, as corrugated as sea shells," but the same material would simultaneously provide both structure and surface, thus eliminating the tedious, costly assembly processes of traditional construction.[43] A team of MIT engineers and architects had already begun designing the first all-plastic house, announced in 1955 and erected in 1957 at Disneyland, appropriately located halfway between Tomorrowland and Fantasyland. Funded by the Monsanto Chemical Company at a cost of about a million dollars, the house was intended as a prototype for a mass-produced version to be sold for an estimated $15 to $30 thousand. As structural engineer Albert G.H. Dietz explained, the Monsanto House of the Future successfully joined "structure and enclosure" in one material, and frankly employed "plastics as plastics rather than as substitutes."[44] Cruciform in plan, the house had four identical eight-by-sixteen-foot rooms cantilevered five feet off the ground from a central utility core containing kitchen and bathroom. Four curved shells of glass-reinforced polyester (or fiberglas) joined to form the continuously flowing C-shaped assembly of ceiling, end wall, and floor of each room, while picture windows filled the two side walls—creating the effect of a set of giant interlocking sculptural television screens. Hidden insulation of polyurethane foam and styrofoam boards complemented the polyester structure, and the interior of the house was finished off with slightly curving walls of vinyl and visionary plastic appliances (including push-button telephones) provided by twelve corporate co-sponsors.

The designers of the Monsanto house blithely expected it to demonstrate the adaptability of mass-produced modular components. One of the earliest publications of the design included alternate floorplan arrangements, accompanied by the claim that "rooms can easily and economically be added or removed to suit the changing space needs of the family."[45] Another noted

43. Douglas Haskell, "In Architecture, Will Atomic Processes Create a New 'Plastic' Order?" *Architectural Forum* 101 (September 1954): 100.

44. As quoted in "Plastics—Shaping Tomorrow's Houses?" *Architectural Record* 120 (August 1956): 210. See also "Here's How They Are Building the First Plastic House," *House and Home* 11 (June 1957): 188, 192; Gladwin Hill, "4 Wings Flow From a Central Axis in All-Plastic 'House of Tomorrow'," *New York Times*, 12 June 1957, p.31; Thomas W. Bush, "Push-Button, Pine-Scented Plastic House With 'Floating' Rooms Shown at Disneyland," *Wall Street Journal*, 13 June 1957, p.5; and "Monsanto Reveals Present and Future of Plastics in Architecture," *Progressive Architecture* 38 (July 1957): 89.

45. "Experimental House in Plastics," *Arts & Architecture* 72 (November 1955): 20–21.

that "the growing or diminishing family can be easily accommodated by the flexible molded modules."[46] Perhaps so. But when the utopian futurism of the Monsanto house finally wore off, some ten years or twenty million visitors later, it took a wrecking company two weeks to bring it down. After failing with ball, blowtorches, chain saws, and jackhammers, the crew attached choker cables and literally tore it to pieces. Thermosetting glass-reinforced polyester had proven too permanent for the sort of adaptability envisioned by engineer Dietz.

Other plastic solutions in architecture proved less permanent. At about the same time Dietz was developing the Monsanto house, a number of entrepreneurs were experimenting with inflatable structures, hemispheric domes of vinyl-coated nylon supported on fan-blown cushions of air maintained at a pressure slightly higher than that outside. Originally developed by Walter W. Bird at Cornell University in 1946, the first inflatables were radomes, lightweight structures that sheltered Arctic tracking stations without blocking or distorting radar waves. After Bird organized a private company in 1956 to promote inflatables for use as temporary warehouses and exhibit halls, several competitors entered the race. Although such applications as vacation homes and temporary covers for all-weather athletic facilities attracted publicity, not much came of the idea.[47]

All the same, the no-nonsense utilitarian inflatables appealed to a generation of architecture and design students who had grown up with plastics and who were being exposed to such nontraditional influences as Buckminster Fuller's geodesic domes, the technocratic modernism of architects Louis Kahn and Paul Rudolph, and the high-tech equipment of the U.S. space program, much of it dependent on high-performance plastics. As they entered upon their careers in the 1960s, they contributed to a "plastic as plastic" aesthetic similar in some ways to that of the 1930s. As before, plastics took on a utopian sheen as materials whose very essence symbolized the future. Although older consumers preferred injection-molded polystyrene or urethane imitations of wood, especially popular for television consoles and "Mediterranean" furniture, younger Americans embraced the "tactile sensuousness" of "Go-Go Vinyl."[48] Taking advantage of the "total design freedom" of plastics, designers offered colorful injection-molded chairs of

46. "Monsanto-MIT Molded Module," *Progressive Architecture* 36 (December 1955): 71, 59.

47. See "Pneumatic Structures," *Architectural Forum* 106 (April 1957): 160–62; "Plastic Bubble Evolves from Warehouse to Airhouse," *Architectural Record* 121 (June 1957): 248; David Allison, "Those Ballooning Air Buildings," *Architectural Forum* 111 (July 1959): 134–39; and "Two Air-Supported Structures for Athletics," *Architectural Record* 135 (March 1964): 209–10.

48. "Go-Go Vinyl," *American Fabrics* 73 (Fall-Winter 1966): 93–94. On the economics of plastics as wood, see Peter H. Prugh, "Against the Grain," *Wall Street Journal,* 21 July 1966, pp.1, 21, and "A Plastic Trend in Furniture's Future," *Business Week* (26 September 1970): 112–13.

smooth polypropylene, beanbags of urethane foam, modular foam cut-outs, and inflatable chairs and couches of transparent vinyl.[49] Combining the buoyancy of inflatables with the stability of thermosetting plastics, Yale architects and students created the "foam dome" by spraying urethane foam over temporarily inflated vinyl forms.[50] The resulting cave-like structures, frequently imitated around the country, seemed paradoxically "as light as meringue, as sturdy as concrete, as sculpturally free as the homeowner and the architect dare to be."[51] Carrying matters to an extreme, the young proprietors of Mass Art Inc., a firm that marketed transparent inflatable chairs, pillows, mattresses, and jewelry, declared their intention of offering for sale an "inflatable fur chair," a "disposable home," and the so-called "Mass Art Man," a life-sized mannequin of translucent plastic.[52] As their marketing director told a reporter, they had "invested a lot of money on the assumption that inflatable furniture will be around a long time."[53] Ironically, of course, it wasn't; it was disposable, and that in fact was the point.

Even *Modern Plastics,* the normally conservative voice of the industry, celebrated in 1968 the centennial of celluloid with a surrealistic projection of "plastics in the 1980s" that emphasized disposability. A group of industrial designers envisioned such innovations as "convenience-oriented" furniture to be discarded with changing styles, entire modular rooms programmed to self-destruct at the end of a decade, zip-out hospital rooms for easy hygienic incineration, and a kitchen-sized vacuum-forming unit for making disposable dishes at home ("just eat . . . throw away . . . and make some more as needed"). According to editor Joel Frados, such future applications of plastics would "smash through all our existing concepts of how we live and work."[54]

Disposables, inflatables, the vinyl "wet look," bright polypropylene moldings, fiberglass saucer houses, foam domes, and artificial environments all came together in an exhibition of "Plastic as Plastic" held at New York's Museum of Contemporary Crafts late in 1968. Installed in a free-form urethane-foam cave, the artifacts and architectural photographs of the exhibit triggered recognition in the popular press of the revolutionary artificiality of plastics.[55] Barbara Plumb celebrated "plastics' coming-of-age" in the *New*

49. Stuart Wood, "When Furniture *Really* Goes to Plastics," *Modern Plastics* 45 (July 1968): 92.
50. Paul Sargent Clark, "Better Living Through Chemistry?" *Industrial Design* 15 (July-August 1968): 33, 35.
51. "New Foam Homes," *American Home* 74 (April 1971): 75.
52. Melissa Hattersly, "Blow-Up," *Interiors* 126 (May 1967): 127.
53. Stuart L. Levine as quoted by Wood, "When Furniture *Really* Goes to Plastics," 91.
54. Joel Frados, "Plastics in the 1980's—A 15-Year Outlook," *Modern Plastics* 45 (November 1968): 128–30.
55. See the catalogue *Plastic as Plastic* (New York: Museum of Contemporary Crafts, 1968) for illustrations of representative artifacts.

York Times, while art critic Hilton Kramer expounded on the "almost Faustian freedom" of "an entire family of materials that can be made to assume virtually any size, shape, form, or color the mind of man may conceive."[56] Periodicals as varied as *Esquire* and *Better Homes and Gardens* covered the exhibit, the latter concluding that plastics inspired "a grace, a technological beauty that may well become the hallmark of a new American culture."[57]

Increasingly, that culture was seen as one of plasticity, of mobility, of change, and of open possibility for people of every economic class. While *Vogue* described the obviously affluent "Space Lib" people of the portable Pneudome, a "shimmering magical environment" that could be inflated anywhere in less than an hour, a design journalist proposed plastic houses as the salvation of riot-torn cities. Future housing projects, he asserted, might be fabricated of mass-produced plastic modules or foamed in place at the site. Completely flexible inside and out, portable, and ultimately disposable, they could be changed as easily as clothing in response to geographical relocation, changing family size, or sheer whim.[58] And, as the designer of an all-plastic Manhattan apartment observed in an ironic echo of the more innocent era of damp-cloth utopianism, his epoxy walls were "easy to maintain—you could hose them down."[59] But perhaps an industry consultant most aptly summed up the meaning of the plastic-as-plastic movement when he declared that plastics best expressed the reality of an "ever-changing . . . fluid . . . expansive" world and enabled people to keep up with it.[60]

The "new" concept of plastics utopianism obviously did not reach all Americans, but the flood of ordinary plastic products did. Bic pens and credit cards, margarine tubs and Big Wheel trikes, vinyl siding and water beds, and scores of other plastic objects, large and small, became commonplace in the 1960s and 1970s, visible evidence of an ability to remake the world over to a degree never before experienced. Not everyone approved. Skeptics, in fact, criticized the same qualities that were praised by the plastic-as-plastic utopians. In 1967, the year of *The Graduate*'s enigmatic commentary on plastics, one journalist noted the "creeping plasticism" of everyday objects,

---

56. Barbara Plumb, "Genuine Plastic," *New York Times Magazine,* 10 November 1968, p.116; Hiton Kramer, " 'Plastic as Plastic': Divided Loyalties, Paradoxical Ambitions," *New York Times,* 1 December 1968, section 2, p.39.

57. "Plastic Moves Uptown," *Esquire* 71 (June 1969): 105–111; Peg Rumely and Nelda Cordts, "The Plastics Explosion in Home Furnishings," *Better Homes and Gardens* 47 (April 1969): 68.

58. "Inflatable, Portable House," *Vogue* 158 (1 August 1971): 117; Paul Clark, "Better Living Through Chemistry?" 33.

59. Victor Lukens as quoted by John Peter, "Victor Lukens Has a Curvey Plastic Pad," *Look* 35 (9 March 1971): 50.

60. Armand G. Winfield, "Excursion into a Plastic Future," *American Institute of Architects Journal* 45 (February 1966): 66.

while another described them as "too adaptable, so glib that they have not yet acquired dignity as a material."[61] At about the same time, "plastic" became a derogatory adjective, meaning fake or phony, used by young people who rejected the explosive materialism of their parents' culture.[62]

As early as 1963, Norman Mailer had used the word "plastic" to characterize the architecture of postwar commercial strips and suburbs, spreading "like the metastases of cancer cells" over the landscape—each "a little more alike" the others, "a little less like anything very definite."[63] By the mid-1970s, Mailer's rhetoric of cancer seemed less metaphorical as the plastics industry experienced one disaster after another. Vinyl chloride monomer, a major ingredient of vinyl, had been found to cause liver cancer in workers who processed it. Consumers had become concerned about carcinogenic and other toxic elements in everyday plastics. Packaging and other so-called "disposable" plastic objects had become a visible component of the garbage overloading landfills and spilling into parks and streets—a kind of environmental cancer. Finally, the oil embargo of 1973 awakened industry and public alike to the dependence of plastics on petroleum. And although only a tiny fraction of the country's oil went to feed the plastics industry, it appeared at times like a cancerous growth, sapping the nation of vital energy needed elsewhere. At the very least, it seemed a tangible sign of the thoughtless, wasteful expansionism that America could no longer afford.[64]

To what extent did plastics create the tone of postwar American life? Obviously there is no final answer to such a question. To the extent that

---

61. Paul Sargent Clark, "The Coming (Any Minute) Revolution for Plastics," *Industrial Design* 14 (October 1967): 65; Walter McQuade, "Encasement Lies in Wait for All of Us," *Architectural Forum* 127 (November 1967): 92.

62. For an early pejorative metaphorical use, see John Perreault, "Plastic Man Strikes," *Art News* 67 (March 1968): 36. The earliest lexicographical reference I have found is by William and Mary Morris, *Dictionary of Word and Phrase Origins,* vol. 3 (New York: Harper & Row, 1971), 217. Although this meaning of the word has apparently all but vanished from normal usage, it is still included in standard dictionaries. The word "plastic" derives originally from the Greek *plassein* ("to mold or form"), which was also used metaphorically to mean "to fabricate, forge, or counterfeit." See Liddell & Scott's *Greek-English Lexicon* (Oxford: Oxford University Press, 1968), 1412.

63. Norman Mailer, "Totalitarianism," in *The Presidential Papers* (New York: Berkley Medallion, 1970; originally published 1963), 177–81. For evidence of Mailer's later opinion of plastics, see Sidney Gross, "Nuts," *Modern Plastics* 48 (March 1971): 51.

64. These issues generated columns of news and commentary, but see especially Paul H. Weaver, "On the Horns of the Vinyl Chloride Dilemma,"*Fortune* 90 (October 1974): 150–53, 200, 202–204; "F.T.C. Accuses Plastic Industry Of Deception on Fire Hazards," *New York Times,* 31 May 1973, pp.1, 26; G.S. Wiberg "Consumer Hazards of Plastics," *Environmental Health Perspectives* 17 (October 1976): 221–25; Sidney Gross, "Garbage," *Modern Plastics* 46 (April 1969): 81; and "Plastics Producers Sort Out the Oil Shortage," *Business Week* (19 January 1974): 84, 86. For a convincing argument that plastics contribute to a substantial net energy savings by replacing materials whose processing is energy intensive, see Christopher Flavin, *The Future of Synthetic Materials: The Petroleum Connection* (Washington, D.C.: Worldwatch Institute, 1980).

possessions reflect and shape experience, plastics no doubt have had an influence in our culture. If in no other way, the familiarity of man-made materials capable of being shaped in an infinity of forms, colors, and textures has predisposed us to accept patently artificial environments as natural and right for us. Paradoxically, this acceptance of the artificial as natural may prove more useful in a stable "solid-state" society like that of the 1980s than in an expansive society like the one that ushered in the age of thermoplastics. The artificial costs less than the natural. And if the artificial is accepted, even when it imitates the natural, then reality need have no more depth or permanence than a movie set.

As far back as 1970, an anonymous architecture critic recognized that two visually distinct interiors of plastic—one a futuristic foamed cave, the other a "Mediterranean" room framed with "hand-hewn" polyurethane beams—both demonstrated an ability to create a "completely abstract, totally synthetic environment."[65] Similar environments, both "futuristic" and "traditional," abound at Walt Disney World in Orlando, once described as "a programed paradise planned by people who improve upon history or nature whenever the need arises."[66] Architectural preservation projects and new shopping centers alike increasingly echo the mock-historical style of Disney World's Main Street USA, with its fanciful gingerbread detailing shaped out of fiberglass-reinforced polyester.[67] Only an enlightened few, guided by postmodernism's artificial reworkings of the past, possess the irony of an art historian who recently declared, after having his Manhattan apartment decorated with imitation marble laminates, that "this is the age of reproduction and it's vulgar and witless to show real materials." But most of us would understand the comment of his wife, who simply said, "I like anything that's fake."[68] We no longer care if what we see is "real," so long as it pleases us and, less consciously, flatters us by reminding us of our ability to shape and control our surroundings, to make of them whatever we desire. The plasticity we impute to the environment derives ultimately from this century's involvement with plastic itself, even though its protean nature and ubiquity render it so ordinary as to be practically invisible.

65. "Plastics: The Future Has Arrived," *Progressive Architecture* 51 (October 1970): 88.

66. Joseph Morgenstern, "What Hath Disney Wrought!" *Newsweek* 78 (18 October 1971): 38.

67. Roland R. MacBride, "In Fire Safety, the Future Is Now," *Modern Plastics* 50 (November 1973): 65. On Disney World see also Margaret J. King, "The New American Muse: Notes on the Amusement/Theme Park," *Journal of Popular Culture* 15 (Summer 1981): 56–62, and Richard V. Francaviglia, "Main Street U.S.A.: A Comparison/Contrast of Streetscapes in Disneyland and Walt Disney World," *Journal of Popular Culture* 15 (Summer 1981): 141–56. For a provocative treatment of the commodification of the landscape, see Robert B. Riley, "Speculations on the New American Landscapes," *Landscape* 24: 3 (1980): 1–9.

68. Quoted by Suzanne Slesin, "Fake Stone Finishes on a Realistic Budget," *New York Times,* 16 July 1981, section C, p.6.

*Popular Culture and Visions of the Future in Space, 1901–2001**

Brian Horrigan

Transportation technologies are particularly susceptible to analysis for their meaning in American culture. We can make something of the way they appear in the popular arts. In those forms, the railroad became a symbol of national unity, efficiency, power, reach; we translated the automobile into an icon of individualism, endless mobility, freedom. Space travel was thus preordained to be freighted with cultural baggage of a still weightier and more emotional sort, with its eschatological overtones, its connotations of departing this mortal coil for the heavens, its promise to unravel scientific mysteries.

Since at least the early 1890s, American popular culture has concluded that the future will be enacted in a single, vast locale—outer space. Removing the future to rocketship interiors and distant galaxies relieved the makers of popular culture of the onerous and problematic task of predicting the future on Earth, a desirable bonus. At the same time, the outer space locale was essentially a *tabula rasa* on which one could freely exercise dreams and fantasies. And in inscribing the tablet, other truths appear between the lines: our values about children and their socialization, our need for heroes and myths, and the notion of our place in the political landscape.

This paper will show some of the ways that meanings inhere in popular cultural forms. It is an attempt to understand how a technology or group of technologies is displayed in American culture. In the case of space travel, the range of display is exceptionally broad, including amusement parks from Coney Island to Disneyland, science fiction "pulps" and comic strips, advertising, children's games, radio programs, movies and Saturday matinée serials, television, and video games. However, I have limited my examination

---

*I presented this paper in a considerably shorter form to the 27th Annual Meeting of the Society of the History of Technology at Cambridge, Massachusetts in November 1984. I am grateful to several friends who read that paper and offered useful criticisms: Joe Corn, Michael Smith, Susan Smulyan, Rosalind Williams, and Larry Levine. Thanks go to my co-participants in this Colloquium, and also to Amy Levine.

here to the space travel theme in the popular media, understanding, of course, that the mass media do not exclusively constitute popular culture. My inquiry is further confined chronologically to the years after World War II, the period of the greatest popular interest in space travel. This particular section of the territory of popular culture had been amply prepared by earlier generations.

Though literate Americans could read space fiction as early as the 1860s (for example, Edward Everett Hale's 1869 story, "The Brick Moon"), it was not until the last decade of the nineteenth century that a truly mass audience was able to find easily what we now recognize as pioneering space fiction. Jules Verne's *From the Earth to the Moon* and *Around the Moon* had been available in English translation, complete with illustrations, as early as the 1870s, but it was the serialized novel in a mass market magazine or newspaper that was to be the most typical means of middle-class discovery of a future in space.

H.G. Wells' *War of the Worlds*—a classic of invasion fiction and not strictly about the future in space—was serialized in an American magazine (*Cosmopolitan*) in 1897, several months before it was available in England in book form. The novel also ran in American newspapers, among them the *Boston Post*. It was there that the young Robert Goddard encountered it, an experience that "gripped my imagination tremendously," as he wrote later in a familiar passage from his autobiographical sketches.[1] A sequel to Wells' popular serial was hastily commissioned from the shameless Garrett P. Serviss; the first installment of *Edison's Conquest of Mars* appeared in the *New York Journal* in January 1898. In this tale, Edison gathers together America's best and brightest scientists to build a spacefleet and a disintegrating ray to defeat the Martians on their own turf. The series was illustrated with early versions of such key icons of space fiction as fleets of space ships and weightless astronauts.

Extraterrestrial themes figured in the very genesis of the "pulp," the magazines of genre fiction printed on cheap paper which became so popular in the twentieth century. Publisher Frank Munsey is credited with the invention of the successful pulp formula, with the transformation of his magazine for juveniles, *The Golden Argosy,* into the adult-oriented *The Argosy* in 1889. Munsey set the standard in the content of these magazines when he selected as *The Argosy*'s inaugural story a translation of André Laurie's *The Conquest of the Moon.*[2]

---

1. Goddard quoted by Tom Crouch, "'To Fly to the Moon': Cosmic Voyages in Fact and Fiction from Lucian to Sputnik," in *Science Fiction and Space Futures, Past and Present,* ed. Eugene Emme, AAS History Series, vol. 5 (1982), 7–25; Frederick I. Ordway, III, "Collecting Literature in the Space and Rocket Fields," Part I, *Space Education* 1:4 (September 1982), 176–182; Part II, 1:6 (October 1983), 279–287.

2. Sam Moskowitz, ed., *Science Fiction by Gaslight* (Cleveland and New York: World, 1968), 28.

The stories of Wells and Verne and their contemporary imitators were redolent with the fear of the new industrial order: of new technologies, of the harnessing of energy, of global explorations and colonization, of the movements of newly-armed nations against each other and against weaker peoples. The popularity of the often terrifying futuristic fiction of the late nineteenth century cannot be understood without the complementary contexts of imperialism and industrialization. These "machine operas" gave ardent, soaring voice to new realities, bringing them onto the stage of popular culture behind a scrim of romance and heroism.

It was merely a matter of time, cash, and chutzpah before someone would stretch these fantasies into a full-bodied, titillating experience. In the sideshows at the Pan-American Exposition in Buffalo in 1901, and a year later at the Steeplechase at Coney Island, visitors could take "A Trip To the Moon," a ride aboard a fanciful spaceship situated within a large circular building. As the ship rocketed, a painted cyclorama moved around it:

. . . once on board the great airship, her huge wings rise and fall, the trip is really begun, and the ship is soon 100 feet in the air. A wonderful, widespread panorama . . . seems to be receding as the ship moves upward . . . . Houses recede from view until the earth fades from sight, while the Moon grows larger, larger. Passing over the Lunar satellite, the barren and desolate nature of its surface is seen . . . . The airship gently settles, the landing made, and the passengers enter the cool caverns of the moon.

Disembarking passengers were greeted by giants and midgets in moon-man costumes, the Man in the Moon on his throne, and dancing moon maidens, passing out bits of green cheese.[3]

In 1903, with Coney's turnstiles turning faster than ever (thanks in part to the new Williamsburg Bridge from the lower East Side of Manhattan) the creators of "A Trip to the Moon" opened a rival park next door to the Steeplechase: Luna Park. The entire phantasmagoric landscape of Luna Park was meant to operate metaphorically as the surface of the Moon, that is, in the words of its mastermind, Frederic Thompson: ". . . a different world, perhaps a nightmare world, where all is bizarre and fantastic." Two years later, Thompson invaded Manhattan, and opened his colossal Hippodrome on Sixth Avenue and 43rd Street. The opening performance in this 5000-seat pleasure palace was "A Yankee Circus on Mars," in which a destitute circus is somehow contrived to land on Mars to entertain the king. But never mind the plot, get to the climax, in which 64 "diving girls" descended a staircase in groups of eight and disappeared into a lake at the foot of the stage. Such a

---

3. The description of "The Trip to the Moon" is taken from *The Official Guide to Coney Island,* published in 1905, and cited in Rem Koolhaas, *Delirious New York* (New York: Oxford University Press, 1976), 32–33; John Kasson also provides an excellent discussion of Luna Park in *Amusing the Million* (New York: Hill and Wang, 1978), 61–63.

description recalls the *gaieté parisienne* flavor of Georges Melies' famous 1902 film, *Un voyage dans la lune* (which had some sporadic distribution in the U.S. around 1905), particularly the chorus line of *jeunes filles* escorting the bearded scientists to the cannon for their shot at the moon.[4]

Places like Luna Park or "A Trip to the Moon" transformed fantasies that were heretofore strictly private or literary into events that were public and real. Paradoxically, these imaginative landscapes left little to the imagination. The event was carefully controlled and organized, the experience essentially passive. Furthermore, one paid for the excursion: sixty million vicarious travelers paid their way into Luna Park in its first five years. Trips to the moon, events on which so many billions of dollars would eventually be spent, neatly slipped into the context of commerce and middle-class hedonism. Space and the Future became a place to escape to for leisure and amusement.

By about 1900, then, the popular image of a future in space had assumed some basic outlines. The future would be first and foremost entertaining; adventure, thrills, and excitement awaited us there. The engines of these adventures would be the spectacular technologies. Conquest was another goal, sometimes interpreted as merely the fact of getting there, sometimes as actual subjugation and colonization of aliens. The future might be fractious: close encounters could be violent, ribald, or amusing. The popularity of space travel themes at sideshows, burlesque houses, and in pulp magazines imprinted the future in space with an indelibly popular, even vulgar character. If the space travel theme ever had the potential to be "serious," genteel, or morally edifying, it was not evident at the beginnings of the century.

As long as popular culture confined its exploration of space to the occasional magazine piece or to bawdy burlesque houses, the subject was bound to remain limited in both audience and theme. That was the case for the first twenty-five years of the century. The picture began to change in 1926, when Hugo Gernsback brought out *Amazing Stories,* the first magazine devoted exclusively to science fiction, or "scientifiction," as he called it. Gernsback was a pioneer in the field of amateur radio, and brought out his first magazine, *Modern Electrics,* in 1908 to cater to the burgeoning interest in educational home hobbies among juveniles and children. He published an occasional piece of futuristic fiction in his *Modern Electrics,* including his own "Ralph 124C 41+" (1911). *Amazing Stories* proved to be quite successful, and by the early 1930s a number of science fiction titles were crowding the westerns, detective rags, and hobby magazines on the pulp racks. Though the science fiction magazines quickly established a separate identity, Gernsback never made strenuous distinctions between the goals of his hobby magazines (such as *Radio News* and *Spare Time Money Making*) and those of his fiction

4. Koolhaas, *Delirious New York,* 78–79.

pulps. Wholesome "scientific" hobbies for juveniles were meant to be complemented by fiction that was sincerely seen as an adjunct to education.[5]

The early science fiction pulps were overwhelmingly devoted to outer space adventures; for all practical purposes, science fiction was space fiction. *Amazing Stories* is credited with inventing the formula that has been identified with science fiction down to the present day, that is, the "space opera," the extraterrestrial epic filled with heroic action, bizarre aliens, and dazzling futuristic technology. This direction was in large part determined by the first editor of *Amazing Stories,* Gernsback's young protégé, David Lasser. Gernsback and Lasser together founded the American Interplanetary Society in the late 1920s, the first American organization devoted to the study of rocketry and the possibilities of space travel, while Lasser's 1931 *The Conquest of Space* was the first serious examination of these subjects in the popular press. Though necessarily speculative—it is hard to imagine what a "technical" book on spaceflight would have looked like in 1931—Lasser's book was a significant departure from the flashy fantasies of the pulps.[6]

Gernsback was a clever and dedicated businessman who survived several reversals in his publishing career, always emerging with his more altruistic motivations intact. He was the first to recognize and exploit the didactic potential of science fiction, the first to understand that the genre could be employed in the service of popular science education. That much of the "science" in this fiction was patently ludicrous hardly mattered, for instruction could also operate on a moral plane. Obvious values such as valor, individualism, and patriotism were subtly joined by such "wholesome" values as the rewards of hard work, ingenuity, and entrepreneurial initiative. Like a Barnum or a Disney on a more modest scale, Gernsback understood that popular culture was something that could be managed and directed, that it could shape popular belief.

Though science fiction in print gained in popularity throughout the 1930s, it was still restricted to the pulps. Rare indeed was the science fiction story in the mass market magazines after about 1920, or the science fiction novel in the 1930s; Olaf Stapledon's reverentially remembered *Last and First Men* and *Star Maker* are two exceptions. As the science fiction readership grew in size, it paradoxically grew in exclusivity. Fans became what sociologist William Sims Bainbridge has identified as the "science fiction subculture." They grew disdainful of the expansion of space fiction themes into

---

5. Paul A. Carter, *The Creation of Tomorrow: Fifty Years of Magazine Science Fiction* (New York: Columbia University Press, 1977), 4–28.

6. Frank H. Winter, *Prelude to the Space Age: The Rocket Societies* (Washington, D.C.: The Smithsonian Institution Press, 1983); David Lasser, *The Conquest of Space* (New York: Penguin, 1931). In addition to Lasser's work, there were several non-fiction treatments of the space theme published in Europe in this period: see Ordway, "Collecting," Part II, 281–282.

newer media, to larger audiences. Nonetheless, from 1930 on, the image of our future in space comes to us primarily through the movies and airwaves.[7]

The successive leaps that the character of Buck Rogers made around the media in the late 1920s and early 1930s neatly summarize this shift. Buck first appeared as "Anthony Rogers" in Philip Nowlan's "sleeper-awakes" story, "Armageddon 2419 A.D." in *Amazing Stories* in 1926. In 1928, he jumped to the pages of the daily and Sunday papers, and in 1932 to the radio and his greatest notoriety. Flash Gordon, a clean-shaven, platinum-haired Buck Rogers imitator in the comic strips, eventually managed to establish a separate fame in the Saturday matinée serials. More than any other characters in popular culture, these two deathless heroes are responsible for the durable image of the future as extraterrestrial, dominated by swashbuckling white males brandishing outlandish weapons against vaguely ethnic aliens, protecting a virginal womanhood and an invincible democracy. Theirs was the world of space opera, certainly not the only available future in American popular culture, but clearly the most ubiquitous.[8]

Space travel in the movies of the 1930s was a limited affair. In both of the major futuristic films of the 1930s—the whimsical musical *Just Imagine* (1930), which takes place in 1980, and the more portentous *Things to Come* (1936), set in the year 2036—space travel is presented as a strictly experimental endeavor of societies with otherwise highly developed technologies. In *Just Imagine,* the hero, "J21," in order to win the hand of his girlfriend, "LN18," sets out on mankind's first interplanetary voyage. His adventures on Mars, ruled by exotically dressed Amazons and menacing eunuchs, are pure burlesque, filled with gags that were at least thirty years old. The dramatic launching of a manned spacecraft forms the climax of H.G. Wells' *Things to Come*. In "Everytown," in 2036 A.D., the first manned spacecraft to the moon is launched from a huge vertically fired cannon, the very device which had sent Jules Verne's voyagers to the moon sixty-five years before! The moonshot is depicted in the film as a controversial decision, the decisive blow against a band of twenty-first-century Luddites. In Wells' script, the exploration of space is equated with scientific progress, indeed, with the very destiny of humanity. "All the universe or nothing," asks Cabal, the progressive leader of Everytown, in the film's stirring finale, gazing out to the stars: "Which shall it be. . . . Which shall it be? Which shall it be?"[9]

7. William Sims Bainbridge, *The Spaceflight Revolution* (New York: Wiley Interscience, 1976), 198–234; Olaf Stapledon, *Last and First Men* (London: Methuen, 1930) and *Star Maker* (London: Methuen, 1937).

8. *The Collected Works of Buck Rogers in the 25th Century,* ed. Robert C. Dille (New York: A&W Publishers, 1977); Robert Lesser, *A Celebration of Comic Art and Memorabilia* (New York: Hawthorn, 1975).

9. H.G. Wells, *Things to Come: A Film Story Based on The Shape of Things to Come* (Boston: Gregg Press, 1975); John Brosnan, *Future Tense: The Cinema of Science Fiction* (New York: St. Martin's Press, 1978), 38–64

Cabal's question was left hanging for most of the decade that followed, a decade wracked by the world war that Wells had so accurately predicted. The two most significant technologies to emerge from the war—the V-2 rocket and the atomic bomb—would, over the course of the next ten years, combine in the popular imagination to create new answers to that question.

In 1945, just a few months after the end of the war, Dick Calkins, one of the creators of the Buck Rogers comic strip, wrote:

> The atomic bomb which the crew of the Enola Gay dumped on Hiroshima opened a future to poor earthbound creatures hitherto undreamed of—except, if I may say so, by Buck Rogers. As Eddie Rickenbacker puts it, "the world is catching up to Buck Rogers."[10]

Calkins' statement, though crude and meretricious, nevertheless reflects contemporary associations between the awesome new technology for world domination, outer space, and science fictional expectations for the future. The government's experimentation with the V-2 and other rockets and the testing of supersonic aircraft confirmed the fancies of earlier generations: the future would be enacted through the spectacle of technology in outer space, and it was upon us.[11]

As if "real" technology was not confirmation enough, there was also paratechnological evidence in great supply, in the form of the unidentified flying object. Beginning around 1947, when journalists coined the term "flying saucer" to describe the "heavenly apparition" of Boise businessman Kenneth Arnold, the UFO phenomenon quickly became confused or conflated with popular expectations about our own future as space travelers. The article by retired Marine Major Donald Keyhoe in *True* magazine in January 1950—which a recent historian has called "one of the most widely read and discussed articles in publishing history"—concluded that the manner in which extraterrestrials observed earth was "similar to American plans for space exploration expected to come into being in the next fifty years."[12]

A full examination of the meaning of the UFO episodes in the postwar years is beyond the scope of this paper, but the phenomenon represents more than a bomb-induced mass psychosis and more than just a virulent outbreak of Cold War paranoia. In popular culture, encounters with aliens or alien

---

10. Dick Calkins, "That Prophetable Guy, Buck Rogers," *Liberty*, 17 November 1945, p. 30.

11. For an excellent discussion of the relationship between the atomic bomb and space travel as perceived in government and the scientific community, see Walter A. McDougall, *...The Heavens and the Earth: A Political History of the Space Age* (New York: Basic Books, 1985), 41–96. See also Frederick I. Ordway and Mitchell Sharpe, *The Rocket Team* (New York: Crowell, 1979).

12. David M. Jacobs, *The UFO Controversy in America* (Bloomington: The Indiana University Press, 1975), 34–107; 57.

intelligence are, with few exceptions (*Invasion of the Body Snatchers* is one), also encounters with their technology. Earthlings constantly remark on how advanced alien civilizations are, that is, how sophisticated their technology seems to be compared to our own, as in the marvelous tour of the underworld city of the Krel in the 1956 space opera, *Forbidden Planet*. Fictional encounters with the alien, even those that occur in a recognizable present, reveal attitudes and expectations about our future in the galaxy. Sometimes the conclusion is triumphant: though we lose most of the public buildings in Washington, D.C., our machines manage to outdistance the aliens in *Earth Versus the Flying Saucers* (1956). Other scenarios are more devastating. An army commander in the 1953 *War of the Worlds,* having just dropped an atomic bomb on a Martian war ship with no discernible effect, despairs: "Guns, tanks, bombs! They're like toys against them!"

To allay these semi-hysterical fears, and to raise public awareness of the potential of a future in space, the growing community of space scientists began to mobilize in the late 1940s and early 1950s. The main event was the First Symposium on Space Travel, held at the Hayden Planetarium in New York in October 1951. The symposium's coordinator was Willy Ley, a German immigrant whose reputation as the nation's preeminent popular interpreter of space science was well-established. Other distinguished speakers included Heinz Haber, an expert in the new science of space medicine, and astronomer Fred Whipple.[13]

The symposium attracted a great deal of public attention. Among the most interested were the editors of *Collier's* magazine, who had been crusading for more access to space subjects for some years. *Collier's* commissioned a series of articles from Wernher von Braun, Willey Ley, and Heinz Haber containing their realistic projections of the future in space. "What you will read here is not science fiction," wrote editor Cornelius Ryan, defensively, "it is serious fact." The articles—which began to appear in March 1952, and continued through 1954—were serious, unsensational, and detailed. Their effectiveness was enhanced by the elegant illustrations of artist Chesley Bonestell, already famous for his astonishing series of planet landscapes published in *Life* in 1949 and his background work for the film *Destination Moon* in 1950.[14]

Simultaneous with these serious treatments of space travel was an unparalleled wave of space travel themes in the media. Pulp fiction boomed again between 1950 and 1953. Space fiction returned to the radio waves after an absence of some years, with programs such as *Dimension X* and *X-Minus One*. Hollywood started an incredible spate of space-science fiction films with

---

13. Shirley Thomas, *Men of Space,* vol. 2 (Philadelphia: Chilton Press, 1961), 213.
14. The *Collier's* series was reprinted in two volumes, both edited by Cornelius Ryan: *Across the Space Frontier* (New York: Viking, 1952) and *Conquest of the Moon* (New York: Viking, 1953).

the "rival" movies, *Destination Moon* and *Rocketship X-M,* which raced each other to the screens in 1950, much as *Star Wars* and *Close Encounters* were to do in 1977. The space opera matinée serial enjoyed renewed popularity, though today there are few who remember *Radar Men from the Moon* and *Zombies of the Stratosphere.* The new medium of television—itself a kind of futuristic dream come true—competed vigorously.[15] This popular culture supernova of the 1950s deserves attention if for no other reason than its sheer volume. What was the character of these forms? What were the goals, beyond entertainment and profit? What messages did the media convey about space, the future, and technology?

The more proximate reality of space technology in the 1950s led, to a certain extent, to a draining away of much of the color and streamlined flamboyance of 1930s space futures. Reasons adduced for exploring space grew more sober, more "scientific," and, as we shall see, more explicitly political. Few treatments of the space theme in popular culture confined themselves to a single rationale, instead combining economic, scientific, and political justifications.

A quite common argument was the simple scientific purpose of exploration. At the surprisingly tragic end of *Rocketship X-M,* all of the surviving crew members of the first excursion to Mars are killed when the ship crashes on re-entry. Nonetheless, the head scientist back at the lab denies that the flight was a failure: "It is proven that interspace travel is not only possible, it is practical. No, the flight of Rocketship X-M is not a failure. Tomorrow we start construction of Rocketship X-M II." The End; roll credits. In *Destination Moon,* a potential backer of the first flight to the moon wants to know: "What's the payoff?" To which the engineer replies: "Dollars and cents? I don't know. I want to do this job *because* I don't know. It's research. It's pioneering. What's the moon? Another North Pole, another South Pole. Why go there? We'll let you know when we get there, we'll tell you when we get back." As it turned out, there really wasn't much more reason to go than just to go, since a fuel miscalculation forces them to leave behind all their equipment and moonrocks. The last line of the movie is a heartfelt "We're going home."

There was also the possibility of "riches beyond your wildest dreams." "Wealth?" asked G.E. Pendray in a popular article in 1946: ". . . there are good—or at least interesting—arguments for the possibility of large deposits of uranium and other radioactive metals on the moon."[16] The crew of

15. Bill Warren, *Keep Watching the Skies!* (Jefferson, North Carolina: McFarland, 1982), Peter Biskind, *Seeing is Believing: How Hollywood Taught Us To Stop Worrying and Love the Fifties* (New York: Pantheon, 1983), 102–105; Bosley Crowther, "Outer Space Comes of Age," *Atlantic* 189 (March 1952): 91–92.

16. G. Edward Pendray, "Next Stop the Moon," *Collier's* 117 (7 September 1946): 11–13.

Rocketship X-M, obviously less altruistic than the scientists back home, remark on the fabulous future of mining on Mars. This notion found its way into children's board games which used space travel as a theme. In Milton Bradley's "Space Game" of 1955, the object of the game is to amass "rare elements" by mining distant planets and stealing from one's opponents. The future in space in such treatments is clearly not one of scientific cooperation among friendly nations, but rather one of conflict and competition between individuals bent on getting rich. The pecuniary motive in popular treatments of the space theme was not new. In Garrett P. Serviss' *The Moon Metal,* published in 1905 in the Munsey pulp, *All-Story Magazine,* a scientist extracts a new metal from the moon which replaces gold as the standard on earth—a story William Jennings Bryan might have liked. In Fritz Lang's 1929 film, *Frau im Mond* (released here variously as *The Girl in the Moon* and *Woman in the Moon*), a financier is persuaded to go to the moon on the basis of a scholarly paper, "Hypothetical Description of Gold Deposits on the Lunar Mountain Ranges."[17]

But the real reason for space exploration, at least for the sages of popular culture, was unambiguous. Less than a year after the atomic explosions that ended World War II, the rocket expert G. Edward Pendray wrote: "So far as a sovereign power is concerned, control of the moon in the interplanetary world of the atomic future could mean military control of our whole portion of the solar system."[18] In 1946, Pendray didn't name names, but any doubt as to what was really meant, in terms of global antagonisms, soon vanished. With the detonation of the first Soviet nuclear test bomb in September 1949, then the Hiss and Rosenberg cases, and finally Truman's public announcement in January 1950 of an accelerated program to develop a hydrogen bomb, it became clear that technological competition with the Soviets would assume the character of a life and death struggle for nuclear superiority. Cornelius Ryan, in his introduction to the *Collier's* series on space travel in 1952, issued an

urgent warning that the U.S. must immediately embark on a long-range development program to secure for the West "space superiority." If we do not, somebody else will. That somebody else will very probably be the Soviet Union . . . Collier's feels justified in asking: "What are we waiting for?"[19]

The fractiousness of our future in space, so pronounced a feature of fictional adventures in space, had rather suddenly assumed real meaning and the

17. For a discussion of children's "space games," see Joseph J. Corn and Brian Horrigan, *Yesterday's Tomorrows: Past Visions of the American Future* (New York: Summit Books, 1984), 28–29; Moskowitz, *Science Fiction by Gaslight,* 36.
18. Pendray, "Next Stop the Moon," 11.
19. "What Are We Waiting For?" *Collier's* 129 (22 March 1952): 23.

58

popular media responded by interpreting the race for space as a Promethean struggle for survival. Space became the final arena in the super-power quest for political domination.

In popular culture the complexities of geopolitics were reduced to a few technological threats. One of the most popular was the moon-based missile launching platform, which became something of an *idée fixe* in the movies of the 1950s. In *Spaceways,* a 1952 release about romance and espionage among space scientists, "Professor Kepler" and an Army general have differing views about the usefulness of an expedition to the Moon:

General: "Have you considered the position, sir, if we lose the race to get there first?"
Professor: "A steppingstone to the moon, to the planets, to whole new worlds!"
General: "And, if necessary, a launching platform for atomic weapons."
Professor: "I hope we should never have to use it for that purpose."
General: "No, but it'll be available if we need it."

In *Riders to the Stars* of 1954, a similar debate occurs between the scientific mastermind of the first space shot and one of the carefully chosen astronaut candidates. "Dr. Stanton" is making his final pitch to the candidates: "If we're not the first nation to solve the problem of space travel, we'll have small chance of survival." The reluctant candidate gets the picture: "Here we go again, boys: wars, killings: every invention seems to have the same end." Dr. Stanton: "The exploration of space by us may be the end of wars! A space platform operated by a dictatorship would make slaves of all free people!"

In *Destination Moon,* released in the tense summer of 1950, the same argument is the decisive factor for the group of captains of industry who have been gathered together in hopes that they will privately finance a moon shot. Though at first skeptical, the group is swayed by retired "General Thayer":

"We're not the only ones who know that the moon can be reached. We're not the only ones planning to go there. The race is on, and we better win it, because there is absolutely no way to stop an attack from outer space. The first country that can use the moon for the launching of missiles will control the earth. That, gentlemen, is the most important military fact of our century."

Needless to say, out came the checkbooks.

That spaceflight and nuclear weapons would combine in the future with awesome consequences was a fact known both to movie Earthlings in the 1950s and to aliens. "Klaatu," the impressively Christ-like extraterrestrial of *The Day the Earth Stood Still* (1950), predicts during his sojourn on Earth: "Soon one of your nations will apply atomic energy to space ships—that will threaten the peace and security of other planets." Klaatu's recommendation was unilateral disarmament—hardly a popular view in the movies or in the geopolitics of the day. On the contrary, received opinion held that our possession of atomic weapons, and hence our power to destroy the globe, *required* that we go into outer space to maintain control. The "higher ground" of outer space, of which so much was heard in the 1950s, was, in fact, synonymous with nuclear domination.

This message, with its overtones of nuclear fear, is implied in nearly every treatment of the space theme in 1950's popular culture. However, the message is often muted by other factors, such as the nature of the audience. This was particularly true of space fiction on television, which, in the 1950s, was directed exclusively to children and juveniles. In the case of the most famous non-fiction TV show on space, Walt Disney's *Man in Space,* the goal was primarily public relations for developing a national space policy. All, however, were framed by the escapist and commercial imperatives of popular media.

The impetus for Disney's first venture into space was provided by the *Collier's* articles of 1952–1953. Among those intrigued by the series was Ward Kimball, a producer at the Disney Studios, who was then casting about for ideas for Walt Disney's first foray into television, a show to be called "Disneyland." The program was intended to advertise the amusement park of the same name, which was under construction. It had already been decided not to build a tawdry, permanent carnival, but something completely new—a "theme park." Further, there was a sense that the park should describe a whole, that it be a kind of empire with a history: a past, a present, and a future. Television programming that would correspond to three areas of the park— Adventureland, Frontierland, and Fantasyland—was easy to imagine, if not already in the Disney vaults. But what about Tomorrowland? By this time, it was probably inevitable that the future's figurative counterpart to Davy Crockett and Tinkerbell would be "Man in Space." As the show developed, it became clear that adventure, fantasy, and frontier heroics would play at least as great a part in the Disney future as they did elsewhere in Disneyland.[20]

Disney asked the *Collier's* authors—Wernher von Braun, Willy Ley, and Heinz Haber—to act as advisers for the new program, to give it shape and direction. Von Braun (a self-described "two-headed monster—scientist and public relations man") was especially eager for the opportunity to publicize his ideas on space travel and technology before the huge TV audience. In addition to their work behind the scenes, von Braun and the others appeared

20. Most of my information on Disney's *Man in Space* was gathered in discussions with Disney's archivist David R. Smith, and from his article, "'They're Following Our Script': Walt Disney's Trip to Tomorrowland," *Future* 1:2 (May 1978): 54–63; on von Braun's media relations, see Erik Bergaust, *Wernher von Braun* (Washington, D.C.: National Space Institute, 1976), 21

in person on the series, which began with the "Man in Space" episode in March 1955. A year later, the second episode, "Man and the Moon," was aired, followed in 1957 by the final episode, "Mars and Beyond." The show combined live action, some fairly dry interviews with the scientists, and animation. The cartoon sections were used primarily to recapitulate the history of flight and rocketry. As Disney himself said during a script meeting: "We are trying to show man's dreams of the future and what he has learned from the past. The history might be a good way to work in a lot of your laughs." And, indeed, here, as elsewhere in the Disney world view, history *is* a lot of laughs, with Archimedes and Robert Goddard momentarily standing in for Goofy and Donald Duck.

The Disney series differed from other contemporary popular treatments of manned space travel in one essential: it was consciously meant to shape public opinion and government policy. While Disney himself may have been influenced largely by the need for programming, the show's advisors were actively engaged in promoting a concerted national space program. "Man in Space" was didactic, even hortatory in tone, confident in its assumption that spaceflight was inevitable, urgent for more action to insure that "a peaceful nation" be first in space. The extent of the program's potential for influencing opinion was understood by many people. As one Los Angeles critic observed, albeit hyperbolically:

"Man in Space" is believable, and Disney has close to 100,000,000 Americans believing. Half of the voting population of the USA has probably reached two impressive conclusions: "It CAN be done!" and "Let's get on with it!"

After the program's first episode aired in March 1955, President Eisenhower personally contacted Disney to borrow the film in order to run it repeatedly for Pentagon brass. The show was aired again in June, and a few weeks later Eisenhower made the historic announcement of America's plans for launching a satellite for the International Geophysical Year. The Disney Studio was jubilant, and wanted to "ballyhoo" the program's next re-run with the assertion of its singular influence on Eisenhower's decision. Von Braun vigorously discouraged such tactics, though the studio's claim probably contained considerable truth.

In spite of the didactic, even moral tone of the program, the Disney show was still commercial television, sponsored at the time by RCA. The commercial message was easily transferred from the television show into the amusement park. On the "Tomorrowland" rocket, visitors came in for repeated aural and visual reminders that TWA was in command. Though the sleek rocket is gone today, one still comes hurtling out of "Space Mountain" into a bizarre and unavoidable encounter with an RCA "Home of the

Future," and Disneyland visitors can ride aboard a—now woefully superannuated—space flight simulation sponsored by McDonnell Douglas.

The addition of corporate sponsorship to the intersection of popular culture and space travel is of more than passing interest, but a detailed examination of the effect of the profit motive is beyond the scope of this paper.[21] A commercial incentive has always been present in the treatment of the space theme in the popular arts. Rides through outer space helped swell Coney Island's coffers; Gernsback's magazines were intended to show a profit, and folded quickly when they did not; *Variety* reported in 1950 that space movies had "good b.o. [box office] potential." Imaginary space travel was (and is) an attractive theme, and people gladly pay money to experience it in a comfortable atmosphere of vicarious enjoyment. Corporations such as RCA and Douglas found it worthwhile, in terms of public image and esteem—if not directly in terms of government contracts—to be the visible sponsors of blissful space events at Disneyland. Space travel as an economic reality (shaped by the nascent aerospace industry) and space travel as popular fantasy (shaped by Disney and other Hollywood mythmakers) developed, as it were, as interlocking spheres on the same fertile soil of Southern California. Nor can it be entirely a coincidence that when Disney turned his empire-building sights eastward, he selected an enormous, swampy site a mere sixty miles from Cape Canaveral.

As the more serious sponsorship of Disney's amusement park rides set them apart from a Coney Island diversion, so also was "Man in Space" intentionally elevated above the then current standard for space fiction television. Disney's series was a family program, with parents watching along with their children, and was to some extent meant as an antidote to the glut of futuristic space adventures aimed at children. Between 1949 and 1956, American kids avidly consumed a TV diet that included "Captain Video," "Tom Corbett, Space Cadet," "Space Patrol," "Rod Brown of the Rocket Rangers," "Rocky Jones, Space Ranger," and "Commando Cody, Sky Marshall of the Universe," to cite only those that aired nationally. Typically, the shows were done live, and aired in fifteen-minute segments several times a week or for a half-hour on Saturday mornings. The acting was exuberant and awful, the plots contrived and recycled from the movies or the pulps, the sets minimal, and the special effects not terribly special (the weekly budget for effects on "Captain Video" was $25.00). Needless to say, they were wildly popular.[22]

---

21. Michael L. Smith examines some of these issues in his essay on advertising and the U.S. space program, "Selling the Moon: The U.S. Manned Space Program and the Triumph of Commodity Scientism," in *The Culture of Consumption: Critical Essays in American History 1880–1980,* ed. Richard Wrightman Fox and T.J. Jackson Lears (New York: Pantheon, 1983), 176–209.

22. Gary Grossman, *Saturday Morning T.V.* (New York: Delacorte, 1981), 135–139.

Captain Video ("Master of Space! Hero of Science!") was the earliest space program on TV, premiering in June of 1949. The Captain was a twenty-second-century intergalactic cop, struggling against future villains and aliens with the aid of a marvelous cardboard arsenal that included an "opticon scillometer" and a "cosmic ray vibrator." The Captain was so popular that at one point he could be seen on the air six days a week, and remained on the air in some form for seven years.

"Tom Corbett, Space Cadet" began in 1950 and had a very successful five-year run. It was inspired by a short novel by Robert Heinlein (whose work was also the source for the film *Destination Moon,* which appeared the same year). Tom and his teenaged cohorts rollicked through the galaxies in the year 2350 A.D. With technical advice supplied by Willy Ley, "Tom Corbett" was meant to be scientifically more accurate than the unintentionally hilarious "Captain Video," though, in watching the show today, one wonders whether Ley really was around very often.

Next to "Tom Corbett," "Space Patrol" is the most fondly-remembered space program of the 1950s. For four years, starting in 1951, kids were beckoned to the tube to "travel into the future" with Buzz Corry, Commander in Chief of the Space Patrol. If the program had an especially military feel, it was probably due to the fact that its creator, Mike Moser, conceived the program during dogfights with the Japanese over the Pacific.

On the whole, earthly political struggles were absent from the scripts of TV space shows. Tom Corbett's twenty-fourth-century planet was peacefully governed by a "Commonwealth of Earth." The Space Patrol operated out of a base on the thirtieth-century artificial planet, "Terra," a member of the "United Planets." Technological advances of the twenty-fifth or thirtieth centuries had led automatically to interplanetary amity and the abolition of war, a common, if naive, prediction in much of the popular culture of the twentieth century. Some programs, on the other hand, attached some strings to the notion of a United Planets. "Rod Brown of the Rocket Rangers," required its Saturday morning fans to take an oath of allegiance to the U.S. Constitution, a kind of weekly loyalty oath for the pajama set.

Though they were for the most part simple-minded entertainments, television's space shows of the 1950s marked a return to Hugo Gernsback's original principle for science fiction, that it was an appropriate way to educate youngsters about contemporary developments in technology. Little lessons in technology cropped up in most episodes of these series, like vitamins sneaked into Malt-o-Meal. On one episode of "Tom Corbett," "Dr. Joan Dale," a professor at the Space Academy, is teaching the cadets about atomic energy, and for the show-and-tell she has wheeled into the classroom "the simplest form of the atomic pile," as she explains, "only a little advanced beyond those used when atomic energy was first developed about 400 years ago." She then explains about graphite rods and controlled reactions, cautioning that "misuse of this pile could cause serious injury." Well, boys will be boys, so when Joan leaves the room. . . "The pile! It's out of control!" Quick fade to the boys' dormitory room, where the mischievous cadet is having his face smeared with "radiation salve." The pedagogy could also be of the moral sort. In an extraordinary episode of "Space Patrol" from 1952, Buzz Corry and Cadet Happy travel back in time to seventeenth-century Salem, Massachusetts, and end up saving a woman from a witchcraft trial. As the episode begins, a narrator intones: "It is 1692. These are dangerous times, for the people of this timeline had given way to mob hysteria as they seek out victims to persecute for the charge of witchcraft." At the height of the McCarthy rampage, the language and setting could hardly have been more pointedly chosen, though the symbolism was probably lost on juvenile audiences.

Adults probably watched some of the series that were on during the evenings, but most of the audiences were made up of children—a fact evident from the advertising. In the blithely surrealistic way of most advertising, commercials showed space heroes of the thirtieth century munching on breakfast cereals of the 1950s, making direct pitches to their audiences to send in boxtops for helmets and rayguns, and exhorting kids to tell Mom to put Ovaltine on their shopping lists. These ads were thus a kind of time travel, which was also the plot device of first and last resort. The future in space collapsed, effortlessly, into the present here in America, or rather Earth.

To children, the commercial message was clear: to be truly like the space heroes of the future one had to buy things they endorsed, or that were part of their regalia. The gimmick of merchandising tie-ins with space heroes was the creation of the John F. Dille Syndicate, which owned the rights to "Buck Rogers." Their licensing of the first Buck Rogers "Rocket Pistol" in 1934 opened the "kiddie-konsumer" floodgates. Macy's sold 5,700 of these toy rayguns in a single day during that Christmas season, no doubt aided by a special "Buck Rogers Train" ride that scooted kids through a futuristic landscape on the store's ninth floor. The *New Yorker* reported that year that a Lynchburg, Virginia department store had replaced Santa Claus with Buck Rogers. (I noted with some pleasure during the Christmas season of 1984 that Gimbels was catching up with its rival, offering "Santa's Space Shuttle," a "multi-media adventure of Christmas-in-the-future"). It has been claimed that there has been more merchandise associated with Buck Rogers than with any other comic character, with the exception of his exact contemporary, Mickey Mouse. During the 1950s, more space heroes meant more stuff. Nearly one hundred manufacturers issued Captain Video products; Tom Corbett's name was on 135 different items, ranging from cheap metal buttons to a pricey playsuit and plastic helmet set.[23]

23. Lesser, *A Celebration of Comic Art*, 37; Corn and Horrigan, *Yesterday's Tomorrows*, 23; 28–29.

Meanwhile, back on Earth, there were grown-up boys playing with grown-up toys. The succession of spectacular space events began in 1957 with Sputnik, and climaxed with the televised landing on the moon in 1969. Many of the space pioneers in the movies and television watched this parade with a mixture of wistfulness, nostalgia, and a sense of *déjà vu.* "Over the years, I saw many things that we dealt with as fantasy come true," said Frankie Thomas, TV's Tom Corbett. "Watching the moon landing on TV, I said aloud, 'Good Lord, that's our stuff' . . . . I felt I had been there so many times before, I could have pointed the way." Science fiction has always claimed the voice of prophecy; the motto of Hugo Gernsback's first pulp, *Amazing Stories,* was "Extravagant Fiction Today; Cold Fact Tomorrow," printed beneath a picture of Jules Verne rising from his tomb. To leap ahead of earthbound science and technology, "to boldly go where no man has gone before," in the famous *Star Trek* split infinitive, became the self-appointed mission of the creators of popular space fiction. Neil Armstrong's first words on the moon unconsciously recalled the parallel moment in *Destination Moon:* "By the grace of God, and in the name of the United States of America, I take possession of this planet on behalf of and for the benefit of all mankind." And on the day of the first Apollo orbit around the moon, Wernher von Braun called Ward Kimball, producer of the Disney space series, and said: "Well, Ward, it looks like they're following our script."[24]

The prophetic gifts of pop culture space heroes faded in importance as the space race commenced. By January 1958, the date of the first successful launch of a U.S.-built missile, science fiction had abandoned the arenas of popular culture where formerly it had commanded huge, rapt audiences. The number of space fiction stories in the surviving pulps peaked in 1953, then dropped rapidly, except for a small boom during the "Sputnik Autumn" of 1957. The movies no longer ventured into space; science fiction cinema was increasingly occupied with blobs and teenage werewolves. Not a single TV show with a space theme was on the national airwaves in 1957. Perhaps, after Sputnik, all of the B-movies and gee-whiz TV shows began to look a little embarrassing. The Soviets had clearly not been expending their space-age intelligence and energy on pop culture. Perhaps audiences simply tired of the repetitious situations in 1950s fiction, and found reality—for the moment—more interesting. Like the western, the "space opera" is a genre with an apparently cyclical nature. So it is surprising that in 1958 no less an authority than *Variety* announced with finality that the science fiction era had ended. Today, inundated by a decade of space odysseys, star wars and star treks, we know that obituary was premature.[25]

At the beginning of this paper I mentioned the remarkable ease with

---

24. Thomas quoted in Grossman, *Saturday Morning T.V.,* 143; von Braun quoted in Smith, " 'They're Following Our Script,' " 63.

25. Bainbridge, *Spaceflight Revolution,* 203–208.

which some technologies slip into the myth-making and romance of American popular culture. Space travel became part of this process, but it is an anomaly—for in this case the technology existed as a fully-embroidered tapestry of the imagination before the fact. Never mind that some of the details—the green cheese, the monsters, and the anti-gravity belts—were a little off. Popular culture transformed the technological imagination, gave it material shape and content, and invested it with politics and consumerism.

Images from popular culture impress us mightily, and have a tenacious hold on our imaginations. Science fiction memories spring forward when we want to conjure up a notion of reality made suddenly, inexplicably fantastic.[26] On the other hand, one can attempt to analyze the motives of the sponsors of popular cultural forms—the entrepreneurs of amusement parks, the producers of movies, magazine publishers, and the like. Their productions may not be "popular" culture at all, but rather "promotional" culture. Unfortunately, this notion implies that modern societies are incapable of developing cultures unsullied by the market, that are truly their own, comparable to those of, say, primitive, pre-literate, or pre-capitalist societies. Certainly there is some reason for insisting that entrepreneurs are the key figures in the framing of popular themes and images. The sponsors themselves, like advertisers, usually demur, insisting that they are merely responding, however shrewdly, to public demands or perceived needs. And entrepreneurs, like the artists or writers they hire, trade freely in ideas and images. As with historic audience analysis, a retrospective critique of sponsors' motives is frustrated by a lack of data, and often results in blanket indictments of elite decision-makers, with mutterings of conspiracy all around.

John Cawelti, in his examination of genre fiction, *Adventure, Mystery and Romance,* writes that when a given story pattern becomes a widely successful formula, it

clearly has some special appeal and significance. It becomes a matter of cultural behavior that calls for explanation along with other cultural patterns.

26. Still, there are unanswered questions about the images of popular culture. How do such images and messages affect us; to what extent do they influence thought or action? Impact theories are especially tempting in analyzing popular science fiction, since its audience is so often a juvenile one. Science fiction author and critic Thomas M. Disch has often asserted—albeit half seriously—that sci-fi is a branch of children's literature. But historians of technology have rarely considered how technological change has affected children, or how attitudes to technology are imparted to the young. Science fiction, as well as other genres of children's literature, such as "boys' books" (Tom Swift, et al.) and hobby magazines, would provide a useful starting place for such an inquiry. We automatically assume that children are more suggestible, that archetypal notions are more directly translated by the media to children than to adults. Though the point cannot be dismissed entirely, it is equally true that popular culture consumers in the past—young or old—cannot be leveled into one, all-seeing, universally responsive audience. Thomas M. Disch, quoted in Peter Nicholls, ed., *The Science Fiction Encyclopedia* (Garden City, New York: Doubleday/Dolphin, 1979), 112.

Cawelti adds immediately that, "unfortunately," the relation between literature and other aspects of culture is an area that remains "rather impenetrable."[27]

It is perhaps, then, most useful (and safest) to think about space fiction as one such pattern of behavior, uniting participants at various moments of twentieth-century culture in a network of associations. The fuel powering the engine of "formula fiction" is imaginative technology. Technology, in other words, is not just a "bias" of these tales of the future; in twentieth-century American culture, it is its reason for being. Buck Rogers and Wilma Deering are not remembered for their teasing romance, but for the marvelous machines with improbably "scientific" names whipped up for them by Dr. Huer. *2001: A Space Odyssey* is in most respects a maverick film, but Stanley Kubrick understood the demands of the genre: he directed his actors into passive roles, so that HAL the computer and the balletic spaceships dominate the picture visually and thematically.

Space fiction is a durable cultural phenomenon. It survives through its steady dependence on established norms and the vitality and optimism of its futuristic promise. There is a negative side, too. Plots of popular space fiction often bear messages from the future that are dismaying, at best—the virulent nativism and racism of the Buck Rogers plots, the harmless beneficence of atomic power in kiddie space series, the shrill anti-Communism in 1950s movies. Such notions were ratified, in a sense, by their proximity to more conventional and wholesome ideals. Yet the popular genre of space adventure, with the centrality and exuberance of its technological theme, has nurtured the machine dreams of successive generations of Americans, and the genre is in turn refreshed by a mass disposition to celebrate and reward the technological imagination.

27. John G. Cawelti, *Adventure, Mystery, and Romance* (Chicago: University of Chicago Press, 1976), 21. See also Herbert J. Gans, *Popular Culture and High Culture* (New York: Basic Books, 1975). Other studies that examine the relationship between technology and popular culture include Leo Marx, *The Machine in the Garden: Technology and the Pastoral Ideal in America* (New York: Oxford University Press, 1964); John Kasson, *Civilizing the Machine: Technology and Republican Values in America, 1776–1900* (New York: Penguin, 1977); John Stilgoe, *Metropolitan Corridor: Railroads and the American Scene* (New Haven: Yale University Press, 1983).

Back to the Future: EPCOT, Camelot, and the History of Technology

Michael L. Smith

The grass is clipped impossibly short and even, like a Marine's haircut. Drop a gum wrapper, and within moments a brightly uniformed groundskeeper sweeps it away. As you walk beneath the eighteen-story high geodesic dome called Spaceship Earth, brass fanfares swell from hidden speakers. Spread before you are the mylar pavilions of Future World, like Greek temples dedicated to the gods of different realms: GM ("World of Motion"), Exxon ("Universe of Energy"), GE, Kodak, Kraft, and AT&T.

Inside each exhibit hall, automated cars glide through intricately animated environments, most of which convey variations on the same message: in the past, inventors and technicians devised quaintly inadequate precursors to current technology, and consumers had to suffer discomforts mildly amusing to the modern viewer. Now, however, all of that has changed. Thanks to sophisticated engineering and free enterprise, "we can fairly closely determine what our future will be," and it will be wonderful. "If we can dream it," disembodied voices assure us, "we can do it."

That, at least, is the message awaiting the millions of visitors to Walt Disney's EPCOT Center in Orlando, Florida. Historians, who are accustomed to somewhat smaller audiences and fewer special effects, have found the relation between technology and culture to be more complex. Some of them have taken EPCOT to task for trivializing and distorting the past.[1] But

---

I am grateful to Bruce Sinclair and to each of the contributors to this volume for fruitful discussions of their topics; to Brian Horrigan for sharing his knowledge of EPCOT; to Greer Hardwicke for suggestions on the history of pageantry; to Joel Kuipers for insights into the ethnology of language; and to Merritt Roe Smith for sharing the unpublished manuscript of his article, "Technology and the Idea of Progress: a Perspective on the American Experience."

1. Elting E. Morison, "What Went Wrong with Disney's World's Fair," *American Heritage* (December 1983): 71–78; Mike Wallace, "Mickey Mouse History: Portraying the Past at Disney World," *Radical History Review* (1985): 33–57. See also Michael Harrington, "To the Disney Station," *Harper's* (January 1979): 35–39, 42–44, 86.

as Brian Horrigan reminds us in these pages, Walt Disney once described his approach to history as "a good way to work in a lot of your laughs," and his audiences have learned to expect as much. Surely historians of technology and a Florida vacation spot can be exempted from each other's search for a usable past.

Yet EPCOT is more than Tomorrowland on steroids. Despite its amusement park format, it functions as a missionary outpost, dispensing a catechism that has direct bearing on the concerns of the authors and readers of this book. In recent years, historians have devoted increasing attention to the social dimensions of technology, asking how a given technology was implemented, how it affected peoples' lives, how it was depicted and perceived, and how attitudes toward it affected its development. By approaching technology as a social process, they have tried to enlist the past more fully as a tool for understanding the future.

EPCOT proposes to do the same thing. Its Future World exhibits constitute the most widely disseminated account of the social history of technology ever assembled. (Along with the adjacent Magic Kingdom, the new park is expected to attract up to twenty million visitors per year.) Whatever we may think of EPCOT's version of the past, the sheer scale of its public exposure invites us to consider it as a case study in the social depiction of technology.

Both the message embedded in EPCOT and the display techniques it employs have important cultural implications. The park's designers, or "imagineers," did not invent the strategies for misrepresentation found at EPCOT; they simply elaborated upon ways of seeing that were already present in American society. But EPCOT's vision of the past, and its use of language and iconography to convey that vision, provide a showcase of impediments that we must overcome in order to see clearly the dynamics of technological change in American culture.

According to EPCOT's promotional literature, the imagineers are dedicated to assembling experimental technologies to create "leading-edge experiences" for their visitors. Why, then, is so much of its presentation devoted to the past? Part of the reason is that EPCOT's notion of the future is itself an historical artifact. It is not by coincidence that the park's layout so strikingly resembles the 1939 New York World's Fair. EPCOT's prime task is to do what that fair set out to do: revive the public's faith in progress, and in technology as the principal agent of that progress.

Faith in progress through technology has contributed greatly to the rhetoric of social cohesiveness in industrializing societies, particularly in America, where its focus on the future helped to compensate for a limited past. Justifiably dazzled by the dramatic social benefits of industrialization, Americans were more reluctant than their European rivals to acknowledge that those benefits were unequally distributed, and that they were accompanied by new problems. But by the late nineteenth century, even devout followers of the gospel of progress had to mine the nation's past more and more selectively to demonstrate a pattern of unqualified improvement. Public criticism of the social implementation of technology tended to elicit two varieties of official response: reformers sought to preserve the social structure by addressing the problems, while true believers experienced a wave of technological fundamentalism, defending the status quo as part of a predestined, inevitable march toward perfection.

The most recent of these fundamentalist revivals occurred in the late 1960s and 70s—the crucial planning years for EPCOT. Responding to widespread criticism from peace and civil rights activists, feminists, consumer activists, and environmentalists, a surprising number of corporate and government leaders shared with their most radical critics the belief that any objections to particular policies were part of a full-scale attack on technology, or on the corporate and government leaders who did so much to implement it. As one member of the Atomic Energy Commission put it in 1970, critics of the nuclear power industry's environmental safety standards were actually undermining "the American philosophy of life."[2]

The "pro-" vs. "anti-technology" debate of the 70s actually had less to do with technology than with the ideological significance that Americans had attached to it. Technological innovations had so routinely been invoked as an emblem of national greatness that any questioning of their efficacy carried ominous symbolic overtones. In such an atmosphere, technological assessment rarely escaped the din of clashing symbols.

The challenge facing technological fundamentalists of the 70s was to modify the gospel of progress to account for public concerns, without sacrificing its millenial vision of technology. The result, as EPCOT attests, is a radical discontinuity between the past and the future. Unlike the humorless perfection of the future, the past is a realm of whimsically flawed but steadily improving conditions, with the present poised uncomfortably in between. Technical innovations succeed each other like a verb being conjugated. Labor struggles, wars (especially the Vietnam War), and the arms race are neatly omitted; environmental issues and energy crises appear as evidence of our problem-solving abilities. Wrenched out of context in this way, events float free of causes and effects and history splinters into nostalgia.

Nostalgia's great appeal is that it suspends events above the gravitational pull of historical processes, inducing cultural weightlessness. In this way, time can be marketed like any other commodity, usually for the purpose of

2. Theos J. Thompson, "Improving the Quality of Life—Can Plowshare Help?" *Symposium on Engineering with Nuclear Explosives, Jan. 14–15, 1970,* Atomic Energy Commission, vol. 1, 1–4.

allowing consumers to revisit their adolescence vicariously. The original Disneyland attracted parents and children alike by serving as a national sanctuary for cultural nostalgia; "golden oldies" radio stations, Trivial Pursuits, and *Back to the Future* carry on this relatively harmless tradition of time travel.

But when nostalgia takes the place of historical perspective, the social cost of severing the past from its context can be great. The very triumphs and breakthroughs that EPCOT celebrates seem routine and insubstantial when removed from the backdrop of roads not taken, conflicting visions, and failed aspirations from which they emerged. As Presidential politics in the 80s has demonstrated, nostalgia elevated to the level of official policy leads to rule by amnesia.

If decontextualized history is nostalgia, decontextualized technology is magic. And magic, however entertaining it may be, works by diverting our attention from the forces and conditions that shape our lives. In Mark Twain's *A Connecticut Yankee in King Arthur's Court,* Hank Morgan finds his knowledge of nineteenth-century technology difficult to implement in sixth-century England, but even more difficult to explain. When he destroys Merlin's tower with a lightning rod, or repairs the holy fountain by installing a hydraulic pump, Hank has no vocabulary for explaining these devices to his audience; nor does he wish to. Instead, he secures his position as Merlin's successor by staging these events as magic, and surrounding them with visual and verbal "effects," such as gunpowder fireworks and polysyllabic incantations.

In twentieth-century America, the gap between technicians and the general public is perhaps less pronounced than the gulf separating Hank Morgan from Arthurian England, but it has been growing. As technology becomes more complex, and changes at an accelerating pace, technical knowledge becomes more specialized and less accessible. At the same time, pressure to encourage that gap has become institutionalized. Advertising has built an industry on Hank's "effects," and governments often find fireworks and incantation more tempting than systematic, public assessment of technology.

The imagineers confronted their own version of Hank's choice: could EPCOT really serve the educational role its publicity claimed, or would it be another Magic Kingdom, assembling words and images that abetted the mystification of technology as a social process? Particularly in the historical evolution of technology as depicted by GE, GM, and AT&T, the visual content of EPCOT's feature exhibits reflects the discontinuity between past and future implicit in their message. For the most part, historical moments are distilled into a succession of brief visual quotes, with only the most immediately recognizable figures and stereotypes warranting inclusion.

Images of the future, on the other hand, appear to have been selected primarily for their visually arresting qualities: exotic settings notable for their abundance of unarticulated space—the desert, the ocean floor, outer space itself; cities composed of huge towers hovering over vast, paved expanses—near-replicas of the futurist cityscapes from the turn of the century.

Like commercials and television newscasts, these images are designed to make the most efficient use of brief exposure, since the audience's motorized armchair view permits only a rapid succession of glimpses. Despite that limitation, it is remarkable how much has been omitted. Other than barrenness and physical isolation, few real changes are apparent in the lavish futurescapes. More troubling is the absence of agency and causality. Technology appears to just happen: choices occur, but only in the marketplace—at the end of the technological process, not at the beginning. Decisions about design and function, social needs and social impact, are rarely glimpsed beneath the shiny, streamlined surfaces. When edited for display value, our view of technology achieves an autonomous, effortless quality only slightly more pronounced at EPCOT than in the rest of American society.

But nothing at EPCOT is simply viewed; language, too, plays an everpresent role. On all of the feature exhibits, we are accompanied by the voices of our PALS (Personal Audio Listening Systems), not really explaining so much as cheerleading ("It's fun to be free," sings the chorus in GM's "World of Motion"; "Energy, there's no living without you," say the voices in Exxon's "Universe of Energy"). A skillful blend of muzak and newspeak, the anonymous, ongoing commentary reflects a broader cultural phenomenon: in matters of technology, we tend to confuse the different functions of language, mistaking invocation for explanation.

Language, anthropologists and linguists remind us, reflects cultural perceptions and priorities. In the Philippines, the Hanunóo people describe their region's plant life with terminology far more detailed than the species classification system applied by Western botanists. Certain Eskimo dialects have evolved a variety of names for the different conditions of snow, rather than a single word for snow itself.

Given the blizzard of technological change that has blanketed the American landscape in recent decades, it is remarkable that the word "technology" has retained any useful meaning at all. Our vocabulary has responded vigorously to technical innovation and change, but more by creating self-enclosed ghettos of jargon than with new tools for explanation. The personal computer alone has introduced a wealth of new terms and new cultural metaphors, and the Pentagon has spoken in a dialect of its own for decades. But when we try to assess the social implications of computers, or automobiles, or nuclear power, we tend to revert to a vocabulary of generalization as opaque as that of EPCOT's PALS, conjuring up visions of an atomistic universe of powerful but unseen forces. Eskimo and Hanunóo observers might be expected to recognize such incantations for what they are: magic worthy of Hank Morgan.

To overcome the distortions of time, space, and language, historians must demonstrate the lived dimensions of technology over time. To paraphrase E.P. Thompson's dictum about class, technology is a relationship, and not a thing. It is not inevitable or autonomous. Every product of technology is an artifact of choices—choices made by certain people, with varying results, and with varying degrees of understanding about the effects of their decisions. Societies adjust to the consequences of those choices according to the power and insight available to them. It is the task of historians to restore our sense of the possibilities and alternatives inherent in the human use of technology.

Engineers and consumers, the two principal species depicted at EPCOT, provide points of common interest for the authors represented in this collection. Bruce Sinclair and Carroll Pursell illuminate key episodes in the emerging social definition of the engineer during the years prior to America's entry into World War I. Brian Horrigan and Jeffrey Meikle explore public attitudes and expectations about technology through two twentieth-century case studies: images of space flight in popular culture, and the development and promotion of plastics. In tracing the shifting cultural identities of their subjects, all four authors shed light on the relation between popular attitudes toward technology and its implementation.

In 1916, the Massachusetts Institute of Technology moved from its cramped quarters in Boston to its current location in nearby Cambridge. In that same year, a civil engineer named J.A.L. Waddell mounted a full-scale campaign to establish a National Academy of Engineers. Sinclair and Pursell present these events as episodes in a larger struggle to evolve a social definition for engineering. The proliferation of engineers at the turn of the century involved not just unprecedented numbers of people, but new varieties of engineering, as chemical and electrical engineers joined the more traditional ranks of mechanical and mining engineers. As David F. Noble has shown, these changes reflected the emergence of powerful new structures of corporate and government patronage.[3] For the generation examined by Sinclair and Pursell, the term "engineering" had acquired so many applications that it no longer described a profession so much as a scramble for technical, institutional, and social identity. As Carroll Pursell notes, engineering had become "contested terrain."

In this context, Sinclair's opening question—Was there an engineering culture?—is particularly intriguing. Before turning to the actual engineers at

---

3. David F. Noble, *America by Design: Science, Technology, and the Rise of Corporate Capitalism* (New York: Alfred Knopf, 1977).

MIT, he notes the arrival of the engineer as a figure in popular culture, embodying attributes that have since attained great familiarity: "straight-talking, hard-working men who dealt in facts, not in appearance or tricky shadings of language"—men with a predilection for the outdoors, a "driving, restless spirit" and a healthy distrust of cities, bureaucracies, and refinements in general.

This image of the engineer as a cowboy with a slide rule has far-reaching implications, not only for the social transformation of engineering, but for our understanding of the relation of gender to technology in modern popular culture. For Sinclair's purposes, this fictional progenitor of Chuck Yaeger, the Marlboro Man, and Buckaroo Bonzai is significant primarily because, like the cowboy, he entered popular culture at precisely the time when he ceased to correspond to the actual experience of those he represented. If engineers felt uncertain that their "different approach to knowledge" would survive their adjustment to emerging corporate needs, their popular image codified what they feared to lose.

At each juncture of the process that Sinclair describes—the debate over merging MIT with Harvard, the decision to relocate MIT's campus, and the competing proposals for the new buildings—the participants were forced to articulate their notions of engineering in unusually frank terms. The issues they raised, and the language they chose, conveyed the uncertainty of a school and a profession in transition: no longer part of the craft tradition or culture of "the mechanic arts" where engineering originated, but not yet fully integrated into scientific research in the service of the new science-based industries and of government. Sinclair captures MIT in the midst of the delicate balancing act required by its transitional status: a social identity, rooted in class-based antipathy for Harvard, but modified by nervous recognition that the emerging prominence of engineering will require refinement of "the stained fingers and rough clothes of the laboratory."[4]

One of the pitfalls of the gospel of progress is its aura of inevitability, leading its adherents to mistake the most successful and influential developments for the entire process. Carroll Pursell's essay on J.A.L. Waddell's campaign to establish a National Academy of Engineers forms an instructive counterpart to Sinclair's discussion, since the Academy proposal failed as completely as MIT succeeded. While Sinclair concentrates on the engineers' efforts to distinguish their professional identity from that of scientists and the guardians of "high culture," Pursell is more concerned with struggles within the ranks of the engineers themselves. The campaign for a national academy

4. Nor was John R. Freeman the last to worry that MIT's graduates would become the "corporals" of industry rather than the captains. In 1985, MIT announced a plan to expand the role of the humanities in its curriculum—an idea prompted, in part, by the distressing tendency of Harvard and Yale's more "cultured" graduates to outpace their MIT counterparts in their ascent of the nation's executive ladders.

moved along a series of shifting fault lines, dividing the engineering community into older vs. younger generation, newer science-based vs. more traditional craft-based branches of engineering, public- vs. private-sector patronage, and even "outdoor" vs. "indoor" tasks.

One of the most telling aspects of Waddell's notion of a national academy was its largely ceremonial function. Like a badge of honor, membership in his proposed organization was to serve primarily as a highly visible, ritualized recognition of esteem. At a time when the engineering community had outpaced its ability to generate symbols of its cultural identity, Waddell hoped that creating the symbol would provide a shortcut to the esteem it claimed to represent. For the most part, interest in Waddell's organization came from those engineers who considered themselves least likely to benefit from the new directions their profession was taking.

This impulse to generate emblems, however, was not limited to those being left behind. The redefinition of the engineer has rarely been so vividly symbolized as in the "Masque of Power," MIT's ceremonial river-crossing pageant, described by Sinclair. On the face of it, this elaborate production—a hybrid of Elizabethan court masque and medieval morality play, staged by architect Ralph Adams Cram—strikes a note of cognitive dissonance. An impassioned critic of modern industrial culture, Cram championed Gothic revival as part of a moral struggle against all of the changes that were carrying MIT to the threshold of eminence. Dressed as Merlin, he personally embodied the antithesis of the engineer.

Yet Cram was right to claim that MIT lacked enduring symbols of its cultural identity. Like the industrial society of which they were a part, the engineers of MIT did not yet recognize the ceremonial significance of their own daily activities. Instead, they borrowed their ritual from an older, largely irrelevant tradition, donning and discarding frames of cultural reference as freely as costumes.

Cram's extravagant production was less an anomaly than it might seem to the modern reader. The engineers' need for ritual definition was only a more pronounced version of a larger cultural phenomenon. For a decade preceding the "Masque of Power," New Englanders had witnessed an increasing use of the pageant as a way of lending a sense of shared significance to community or institutional events. The next step in the genealogy of the pageant closely followed the engineer's progress. David Nye's *Image Worlds,* an excellent analysis of General Electric's photographic archives, describes the great significance assigned to similar pageants and skits presented by engineers and managers at GE's prestigious summer camps.[5] Beginning in the

---

5. David E. Nye, *Image Worlds: Corporate Identities at General Electric* (Cambridge, Mass.: MIT Press, 1985). For a thinly fictionalized account of GE's summer camp pageants in the 50s, see Kurt Vonnegut, *Player Piano* (1952; repr. New York: Delacorte, 1977).

1920s, GE invited high-level employees to annual summer retreats, where they enacted corporate concerns and goals while costumed as medieval knights, Shakespearean actors, Roman soldiers, and the like. If there was an engineering culture, it dovetailed neatly with corporate culture, where the need for signifying rituals was even more acute.

If popular images of the engineer lagged behind the pace of historical change, caricatures of space flight and plastics began by leaping ahead of events, anticipating marvels just past the horizon. Brian Horrigan examines depictions of space flight in twentieth-century American popular culture, tracing the progress of a concept relatively unhindered by the limitations of actual technical developments. Like aviation, space flight enjoyed a lengthy "prehistory," flourishing in the popular imagination long before the first rocket left its pad.[6]

Whether in the pulp magazines, film, or television, science fiction since the 1920s has returned so frequently to outer space that, as Horrigan notes, "for all practical purposes, science fiction was space fiction." The tendency to equate space flight with science, however, has not been limited to Buck Rogers and *Amazing Stories*. American response to the launching of the first Sputnik in 1957 is a case in point. Newspapers and magazines suddenly hired "science editors," presumably to cover all aspects of science and technology, but primarily to report space news. Placing a satellite into orbit was an engineering problem of no great complexity by 1957; but in response to Sputnik, the nation's educators received a Congressional mandate to accelerate the entire science curriculum in the public schools. Somehow space flight had become the preeminent popular emblem for science.

The characteristics of that emblem are noteworthy. Long after it was apparent that space travel would require resources far beyond the capacity of any one individual, science fiction and the popular media continued to depict the conquest of space as a task for the single explorer. Like Captain Video before them ("Master of Space! Hero of Science!"), American astronauts were depicted as lone heroes, combining the skills of scientist, inventor, and pilot—a slightly elaborated version of Sinclair's fictional engineer. Perhaps one reason for this persistence can be found in Horrigan's discussion of the appeal the image had for children and adolescents, at whom much science fiction was aimed. But it may be that space travel, with its overtones of both conquest and escape, served as an imaginary antidote for the sense of lost autonomy experienced by recent arrivals in the corporate hierarchy—including the engineers and scientists who in their adolescence made up such a large part of the science fiction audience.

---

6. Joseph J. Corn, *The Winged Gospel: America's Romance with Aviation, 1900–1950* (New York: Oxford Univ. Press, 1983). See also Joseph J. Corn and Brian Horrigan, *Yesterday's Tomorrows: Past Visions of the American Future* (New York: Summit Books 1984).

Jeffrey Meikle's essay on plastics appears, at first glance, to view technology's popular culture from the opposite end of Horrigan's telescope, drawing us from the farthest reaches of the galaxy to the mundane realm of Tupperware and Big Wheel trikes. Like Horrigan, however, Meikle demonstrates the important relation between technology and the popular expectations surrounding it. Meikle's task is an ambitious one: to examine the technical developments that launched the plastics industry alongside the shifting applications of plastics as a cultural metaphor. One of the most intriguing aspects of his findings is the way in which the attitudes of Americans toward plastics mirrored their more general tendency to shift notions of technology from one extreme to the other, and to equate technological development with larger cultural values.

Plastics were only one among many technological innovations that inspired initial expectations of sweeping social transformation. At the turn of the century, American newspapers hailed electric lights as harbingers of a crime-free era. In the 30s, the airplane won acclaim as the catalyst for international unity, since it would eradicate cultural distances and render war too horrible to contemplate. After the first Soviet atomic bomb test in 1949, U.S. advocates of the hydrogen bomb praised it as the weapon that, by its unthinkable destructiveness, would prevent a nuclear arms race. And in the 50s and early 60s, the proponents of nuclear power equated it with unlimited cheap energy, universal affluence, and a cleaner environment. Having little direct access to most of these technologies, or to the decision-making process they represented, the public based its expectations on the early promotional efforts accompanying each development. Given the nature of these efforts, these expectations were bound to be disappointed, and the stage was set for pendulum swings in popular attitudes toward technology.

What made plastics different, Meikle observes, was that their great versatility as a substance gave them resilience as an emblem of social values. Particularly after the adoption of thermoplastics, the ways in which people experienced plastics in their daily lives revealed some of the characteristics of society. Innovation vs. imitation, durable vs. throwaway, economic boon vs. environmental hazard—plastics seemed to accentuate the alternatives available to industrial America, and to embody its strengths and weaknesses. Seen in this light, the popular redefinition of the word "plastic" from "resilient" to "phony" in the 1960s, for example, raises questions about the relation of technology and culture that far exceed the function of metaphor. Both the development of plastics and the social perception of their emblematic significance evolved and interacted so pervasively that Meikle's essay is really a skillfully developed introduction to a much larger study.

At the risk of casting these essays in a role their authors never intended, we might imagine them as the basis for a supplemental EPCOT exhibit. Lacking a corporate sponsor, this pavilion could turn instead to some appropriate historical figure for its god of the realm. In light of his penetrating critique of early consumer culture, as well as his abiding interest in the reforming capacity of engineers, Thorstein Veblen is ideally suited for that purpose. The obvious choice for a "theme ride" in Veblen World is an adaptation of Cram's "Masque of Power": passengers will join MIT engineers on their barge as it crosses the Charles and enters the twentieth century; Mark Twain, who knows a great deal about rafts and rivers, will stand in for Cram, but instead of dressing as Merlin, Twain will play the part of Hank Morgan. From the receding shoreline, J.A.L. Waddell and his colleagues will gesticulate angrily, protesting that they should be piloting the raft. And on the approaching shore, the engineers and visitors will find Leo Baekeland and Wernher von Braun poised to carry them by thermoplastic rocket to the Forbidden Planet.

Recent scholarship could provide a basis for several other new pavilions, each with its appropriate historical figure presiding: the Charlotte Perkins Gilman Hall of Gender and Technology; Big Bill Heywood's Wide World of Workers; the W.E.B. DuBois Showcase of Race in Industrializing America; and the C. Wright Mills Power Circus, featuring a three-ring display of the shaping of technology by the state, the corporations, and the military.

A final exhibit would draw its inspiration from *The Gods Must Be Crazy,* a South African film currently enjoying great popularity in the United States. Like *A Connecticut Yankee,* the film is organized around the sudden appearance of some aspect of modern society into a pre-industrial culture. In *The Gods Must Be Crazy,* that culture is an African village rather than Camelot; instead of Hank Morgan, the visitor from the industrialized world is a Coca-Cola bottle discarded from a passing helicopter. We laugh, a bit nervously, at the villagers' efforts to find a function for this gift from the gods, and to fit it into their belief system. Then, in a sequence that Veblen would recognize, one of the villagers realizes that the object's singular uselessness has introduced concepts of property and envy into their community, and he sets out to return it to its proper realm.

In order to translate the villagers' perspective into our own realm of experience, one final exhibit consists of an enormous empty room. Through 3-D film and holographic projections, visitors receive the impression that all the products of our applications of technology are raining down on them from the sky: electric toothbrushes, microwave ovens, modems, radar detectors, tanning lamps, garage door openers, food processors, and MX missiles. Just before each object reaches the ground, it turns into a Coca-Cola bottle. As the bottles accumulate, the audience slowly realizes that they are forming a structure: a vast cathedral built entirely from Coke bottles. The technical sophistication of the projected images is so convincing that the audience cannot distinguish illusion from the real thing.

CONTRIBUTORS

Bruce Sinclair is a Professor in the Institute for the History and Philosophy of Science and Technology, at the University of Toronto.

Carroll Pursell is a Professor of History at the University of California, Santa Barbara.

Jeffrey Meikle is an Associate Professor in the American Studies and American Civilization Programs at the University of Texas at Austin.

Brian Horrigan is an independent scholar living in Washington, D.C.

Michael Smith is an Associate Professor of History at the University of California, Davis.